The New Telecommunications

Series in Communication Technology and Society
Everett M. Rogers and Frederick Williams, Editors

Everett M. Rogers, Communication Technology: The New Media in Society (1986)

Frederick Williams, Ronald E. Rice, and Everett M. Rogers, Research Methods and the New Media (1988)

Robert Johansen, Groupware: Computer Support for Business Teams (1988)

The New Telecommunications

Infrastructure for the Information Age

———————————— ∎ ————————————

FREDERICK WILLIAMS

THE FREE PRESS
A Division of Macmillan, Inc.
NEW YORK

Maxwell Macmillan Canada
TORONTO

Maxwell Macmillan International
NEW YORK   OXFORD   SINGAPORE   SYDNEY

The Free Press
A Division of Macmillan, Inc.
866 Third Avenue, New York, N.Y. 10022

Maxwell Macmillan Canada, Inc.
1200 Eglinton Avenue East
Suite 200
Don Mills, Ontario M3C 3N1

Macmillan, Inc. is part of the Maxwell Communication Group
of Companies.

Printed in the United States of America

printing number
1   2   3   4   5   6   7   8   9   10

Library of Congress Cataloging-in-Publication Data

Williams, Frederick
     The new telecommunications : infrastructure for the information
age / Frederick Williams.
          p.   cm.—(Series in communication technology and society)
     Includes bibliographical references and index.
     ISBN 0-02-935281-9
     1. Telecommunication.   I. Title.   II. Series.
TK5101.W458   1991
384—dc20   91-11635
                                                            CIP

$ 35.00

Telecommunications and the "coming intelligent network" will provide the next strategic advantage in a wide range of business and public service environments, contends Frederick Williams, a nationally known authority on communication technology. In the last decade, he shows, telecommunications has moved from a support role as a utility to a new infrastructure essential for competitive advantage in business, increased productivity in public services, and economic development in cities, states, and nations.

Telephone companies must take the lead, Williams argues, in promoting the use of telecommunications as a developmental tool. He provides numerous examples drawn from field research on innovative applications of the technology and how it adds value in large and small businesses, education, health, scientific research, and residential services. Williams also illustrates how telecommunications can support new patterns of business decentralization, community development, and revitalization of the rural economy.

Williams foresees the increasing coalescence of computing and telecommunications as a "coming intelligent network" able to combine traditional and new services on a common high-performance network. "Communications intensive" companies, he shows, have increased productivity and customer satisfaction with the "value added" from advanced systems. Williams explores the consequences of worldwide deregulation and privatization and looks ahead to the likely future for U.S. telecommunications. He proposes that providers, especially the "Baby Bells" and other local

(Continued on back flap)

Contents

Preface

In the last decade, telecommunications has moved from a background role of a utility to applications meant to create new competitive advantages in businesses, increased productivity in public services, and economic development in cities, states, or nations. Whether we consider voice or data networks, broadcast and cable services, new business links with suppliers, bankers, customers or clients, or special high-capacity urban, state, national, or global networks, these new applications are moving telecommunications from the realm of "overhead" to that of "strategic investment."

The key theme of the New Telecommunications is to identify innovative applications in a wide range of business, public service, and residential environments, as well as to see how new telecommunications services are an important infrastructure component in city, state, and national planning. What are the new telecommunications applications? How do they create value? What are the opportunities for strategic investment? How can telecommunications investments be evaluated? And what is the likely future for U.S. telecommunications now that the divestiture of AT&T is behind us?

ACKNOWLEDGMENTS

In the background research leading to this volume, I am especially indebted to my former University of Texas colleague, Heather Hudson (now of the University of San Francisco), and to current colleagues Jurgen Schmandt, Robert Wilson, and Sharon Strover, plus graduate students Harmeet Sawhney, Jill Ehrlich, Liching Sung, Richard Cutler, Joan Stuller, and Eloise Brackenridge. Encouragement and support from the College of

Communication and the Lyndon Baines Johnson School of Public Affairs at the University of Texas at Austin were deeply appreciated, as were research funds from such sources as Southwestern Bell, AT&T Communications, GTE Southwest, The Texas Telephone Association, the Ford Foundation, The John and Mary Markle Foundation, and the Aspen Institute for the Humanities. For time and assistance in completing the final touches while in residence as a Senior Fellow I am grateful to the Gannett Foundation Media Center at Columbia University.

∎

ONE

On the New
Telecommunications

As we witness the benefits of the coalescence of telecommunications and computing, many are calling this the arrival of the *intelligent network*, which is as much a national information resource as a communications medium.

1

∎

The Coming
Intelligent Network

An optical scan records slight movement on a near $-273\,°C$ icy plain. The image is digitized and transmitted some 2.8 billion miles to be decoded by scientists who see the first example of a nonterrestrial ice volcano. It erupts on Triton, the first moon of Neptune.

Scholars estimate that 2 billion viewers witnessed the opening ceremonies of the 1988 Seoul Olympiad, with an additional billion seeing later news reports—by far the largest audience of a single event in the history of humankind.

Japanese planners contemplate a 19-city "technopolis" chain of newly constructed science cities. They will be linked by a fiber optic backbone offering services ranging from supercomputer access to global television entertainment; in perhaps 20 years the network will offer computer-assisted "all-voice" language translation service.

European Common Market representatives declare telecommunications to be a service for economic development, rather than only a public utility. Plans are initiated for all-European network services, including a world videotex standard.

Glued to his Apollo workstation, Arun Sharma of Bangalore, India, debugs a glich in software used for manufacturing VLSI chips. He "hands" his solution over to Tom Burke, the production supervisor, who gives feedback so Arun can finish the job. They work as a team although Arun is separated by 14,900 kilometers from Tom in Dallas, Texas.

Peter W. is anxious; it is stormy, midnight, and his mother has not yet returned from her meeting, so he dials her. Slowed by a rainstorm on the Garden State Parkway, Mrs. W. responds by cellular car phone that she will be home in 30 minutes, and for Peter to get under his covers and not to worry. He says he's already under the covers with his cordless phone.

The foregoing are but snapshots in the expanding world of the new telecommunications. The increasingly "intelligent" network is bringing us new advances in sciences, as it is in ways of doing business and the businesses that we do. The network is the key to new efficiencies and expansion of public services. And living on the intelligent network promises new quality and opportunities for residential life. In this introductory chapter, we examine the rapidly expanding world of the new telecommunications and what it means for life as we cross the cusp into the twenty-first century.

■   ■   ■

THE EVOLUTION OF MODERN TELECOMMUNICATIONS

From the Speed of Transportation to the Speed of Light

Beginning roughly in 1819 when Danish physicist Hans Christian Oersted speculated that deflecting a needle by energizing a magnet might serve as a signaling device, continuing in the 1820s with French physicist André-Marie Ampère's invention of the astatic needle, and culminating in 1844 with American inventor Samuel Finley Breese Morse's demonstration for the U.S. Congress of an electromagnetic "telegraph," the age of modern telecommunications came into being. In this age, the speed of transporting human messages, with a few exceptions, leaped from the limits of messages carried by the transportation system (10 to 50 miles per hour including carrier pigeons) to the speed of light (186,000 miles per second).

Telecommunications, in essence, is the application of technology to extend the distance over which humans or their intermediaries, such as computers, can communicate. The "carrier" is an energy form rather than the medium in which the message is stored when it is moved by the transportation system. In public use, the term usually refers to the telephone, telegraph, or data networks as point-to-point communication, or radio and television as broadcast communication. In more technical use, telecommunications involves the modulation of an energy form onto which we impose analog or digital patterns representing our messages, sounds, images, or algorithms. While such ancient forms of communication as smoke or fire signals, drum beats, or semaphore flags could fit within the broad definition of telecommunications, modern applications are based upon modulation of electrical, electromagnetic, microelectronic, or light energy.

From "Wireless" to Broadcast Networks

Modern telecommunications encompasses applications ranging from point-to-point data or voice communications to international television broadcasting, as contrasted with an early history of mainly point-to-point uses, or essentially the telegraph. Broadcasting initially evolved from attempts in the 1860s to create a wireless telegraph. Theories of the existence and nature of electromagnetic radiation gave way to experimentation in signaling. The first patent of a practical wireless telegraph is associated with Guglielmo Marconi, who in 1895 demonstrated the sending of long-wave radio signals over a distance of one mile. By 1901 this distance extended to a transatlantic scale.

Early in this century, broadcast technology improved in range and fidelity, but the idea of using it beyond telegraphy did not occur until the end of the first decade when it became evident that one might send messages to any of a group of receivers, or an audience, so to speak. The example often cited in U.S. literature on this milestone was a young wireless telegraph operator who monitored the sinking of the *Titanic* and rebroadcast it to an audience of other operators in the New York City area. (This operator was David Sarnoff who went on to found the National Broadcasting Company.) By the 1920s radio receivers were available to the public and by 1924 a 22-station "network" carried a coast-to-coast hookup of a speech by President Calvin Coolidge. Networks in popular broadcasting began and continued as the linking of individual broadcast stations.

Demonstrations in several countries of broadcasting moving images by telecommunications began when radio was on its way to becoming a popular medium in the 1920s. This new service, combining images and sounds, was called "television" in a *New York Times* article of April 8, 1927, reporting on a demonstration staged by the American Telephone and Telegraph Company. (The article also cited that any "commercial use was in doubt.") Television services eventually paralleled radio in the evolution of interconnecting stations by networks.

In most countries of the world, radio and television networks were developed as national services, operated by the government or some type of public organization. In the United States, history took a slightly different turn in that broadcasting was developed along commercial lines, with government jurisdiction mainly focused upon keeping service areas from interfering with one another. Today, we are seeing some blurring of these distinctions. Many countries have or are introducing commercial systems to complement their public services, and the United States has given attention to the development of "public broadcasting" services, including "citizen access" to some services. Broadcasting has also been substantially influenced by advancing technologies, including coaxial cable and satellite distribution, eventually to be complemented by fiber optic transmission. As we shall see, these new distribution systems are growing similar to point-to-point systems, so that eventually the same technical networks may serve both.

Point-to-point wireless telecommunications has flourished in the forms of radio-telephone, paging systems, and the newly expanding cellular mobile telephone. It also contributes important links and backups for the world's terrestrial networks, a point revisited many times in this book. A remarkable illustration of advances in these areas was that telecommunications allowed the world to witness man's first step onto the lunar surface, a communication miracle as much as a space technology one.

From Point-to-Point to Switched Networks

Point-to-point communications, as described earlier, grew first with telegraphy, then in the 1870s with the invention of the telephone. Space limits our describing differences of opinion over who exactly developed the first telephone, although in the United States the patent trail favors the independent work of Elisha Gray and Alexander Graham Bell. Both Gray and Bell were exploring the possibility of sending multiple messages on a single telegraph circuit by using frequency separations. Gray, a trained and disciplined scientist, saw the possibility of using this approach to send crude voice messages, but stuck to his objective of a better telegraph. Bell, a teacher of the deaf and a tinkerer, seeing the same voice transmission possibilities, moved in the new direction, and probably most importantly, beat Gray to the patent office in 1876. By the 1880s, the telephone was well on its way in business-office uses, and soon to become a residential item.

Figure 1.1 illustrates the evolution of point-to-point telephone circuits into a switched network. As additional users are added to the system, the need for connections grows exponentially, until one reaches practical limits on the number of linkages. The problem is solved by connecting users to a central office, called a "switch" in telephone jargon because lines are switched open or closed with one another. The early switch was the

Figure *1.1*. From Point-to-Point to Switched Networks

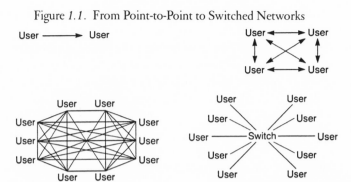

"switchboard," where the famed lady operator literally plugged different lines into one another, given the request of the caller. This function was taken over by cleverly designed mechanical, then electromechanical devices. Today's switch is literally a computer that not only makes the connections but routes calls over the most efficient circuits, records billing information, and helps to troubleshoot the network if there are problems.

Until the divestiture of American Telephone and Telegraph in 1984, the U.S. network was a monolithic hierarchy of switches. Every telephone was connected to a local switch, typically called the "local exchange." Local exchange switches, in turn, were connected by "interexchange" networks, including access to national and international long-distance networks. If you were in an office building, there might be a smaller, internal network served by your own on-premises operator, or switch, this typically called a "private branch exchange," or "PBX" for short. (More advanced on-site switches have been called "PABXs," or private automatic branch exchanges.)

Despite the fact that telecommunications has often been defined in terms of point-to-point or multipoint links, the most modern technical as well as metaphorical manifestation is in terms of network, an interlinkage of participating communicators, often interactive (i.e., two-way), and increasingly "intelligent" in the sense of computer-based self-management as well as information and service capabilities. Already the intelligence provided by computers can offer further services to the users, including least-cost call routing, record keeping, interfacing of different computer systems, and someday, language translation.

A Three-Dimensional View of Applications and Services

Modern telecommunications can be examined along several lines. One approach, illustrated in Figure 1.2, makes distinctions along three dimen-

Figure *1.2*. Three-Dimensional Model

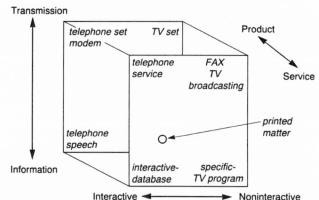

sions: (1) whether the activity is one of simply providing transmissions as contrasted with providing information, (2) whether we are considering a product as against a service, and (3) whether the activity is interactive.

Given the multiple dimensionality of the model, it is possible that some particular products and services will share certain qualities, and this can be illustrated. For example, a television set can be differentiated from television broadcasting as a distinction between a "product" and a "service"; but both relate to "transmission" and are typically "noninteractive." Television broadcasting, by contrast, is distinguished from a specific program as the latter reflects the providing of "information" rather than a "transmission" service. Videotex can be distinguished as "transmission," a "service," and "interactive." The purpose of these distinctions is not so much to create a conceptual model, but to illustrate the parameters to be studied in investigations of network services and products. As we shall discuss, "interactivity" is an important characteristic of modern telecommunications.

In everyday work with telecommunications, one encounters a myriad of technology, systems, and services. Table 1.1 summaries major examples of these, most of which can be located in the three-dimensional model.

NOTES ON THE "INTELLIGENT" NETWORK

Coalescence of Computing and Telecommunications

As telecommunications and computing progress and coalesce, they are blending into a single system which many are now calling the "intelligent network." Just as machines were the principal technology of an industrial

Table *1.1.* Examples of Current Systems and Services

Cellular Mobile. Cellular telecommunications systems include a network of low-power broadcast grids where every cell has a different frequency so that many callers can share the overall network. As callers travel from one cell to another, their communication channel is automatically shifted to one of the different frequencies.

Centrex. In many major telephone service markets, the local exchange company can provide specialized switching services for connecting telephones (like an "intercom") within a company or organization, linking telephones to outside lines or long-distance services. This type of service replaces the need for the company to operate its own on-site switchboard (PBX).

Coaxial Cable. High-capacity cable allows for the simultaneous transmission of many individual messages including moving video images which require large bandwidth. Cable is known for its application in the distribution of television signals to homes. However, modern cable has many enhancements, including two-way or interactive transmission, simultaneous delivery of voice, data, and images, and applications ranging from security systems to remote reading of home electric or gas meters.

Communications Satellites. These are broadcast relay stations that because of their position above the earth can disseminate signals over a wider area than a land-based station. When in an orbit matching the movement of the earth's surface ("geo-synchronous"), earth stations can easily "lock on" to the satellite and need not incorporate expensive tracking mechanisms. As satellites become able to broadcast increasingly powerful signals, earth stations can be reduced in size, making satellite communication much less expensive and more widely available.

Facsimile. FAX allows for the transmission of images rather than individual characters of documents, graphs, or photographs. Depending upon the level of service required, different speed and fidelity specifications are available. Such service allows for the transmission of entire reports, along with graphics, illustrated product catalogs, sketches, contracts, and the like. FAX machines can be set for delayed operation during nighttime hours.

High-Performance Networks. Networks have and are being developed for the inter-connection of major computer centers, including supercomputers. Mostly, these are government networks, the best-known of which is NSFNet.

Integrated Systems Digital Network. ISDN is a newly available service in some parts of the country where one can combine two data or voice channels and one lower-capacity signaling channel on a single telephone circuit. This means, for example, that with a single telephone call, you could link two computers over one channel, have a voice conversation on another, and exchange a slower-speed FAX image on the third.

Local Area Network. LANs are dedicated communications networks often used to link individuals in a building, buildings in a complex, or the geographically

Table *1.1.* Examples of Current Systems and Services (*cont'd*)

separated units of a single company or organization. Such networks can be provided by the local telephone company as a privately leased service, or leased or installed by other telecommunications providers. One popular application is an installation meant to serve tenants in an office building or industrial part, usually called "shared tenant services." LANs developed specifically for computer linkage are typically of sufficiently high speed and capacity so that computers can interact with a minimum of delay.

Microwave Relay. These relay systems allow for the line-of-sight transmission of many simultaneous voice, data, and image signals from tower to tower. As a substitute for wired telecommunication systems, these relay stations have greatly reduced the cost of building telecommunications networks.

Optical Transmission. These systems involve the modulation of light waves as a communications carrier. Two common forms are highly focused "laser" beams and optical fibers that serve as communications channels. Optical systems have the potential to be far less expensive than traditional telecommunications systems. They have a large signal capacity, and are freer from interference than electrical or electronic systems. Much of the planned expansion of the public switched (telephone) network will use optical fiber technology.

Paging. Sometimes called "beeper" services, these allow contact to an individual anywhere within the range of the broadcast signals. The least expensive of the services signal an individual who can then respond with a telephone call. More sophisticated and expensive systems can receive alpha-numeric messages (i.e., text with letters and numbers).

Value-Added Networks. These networks use various combinations of the preceding transmission channels to provide reliable, less expensive telecommunications service. "Value-added" refers to the offering of processing capabilities such as storage and forwarding of messages at a later time, least cost routing, error checking, and detailed accounting records. A particularly significant characteristic of such networks is their use of packet switching, which breaks up a message into small packets, sends each one along the fastest, cheapest, or most reliable route, and reassembles the packets at the destination computer.

enterprise, the intelligent network is the chief tool of the information age. Modern investments in the intelligent network can enhance industrial productivity as well as increase the breadth and efficiency of human services (e.g., education, health, and security).

Transmission and Switching

The intelligent network is more than a communication technology; it is an information "resource," and broadly speaking, an infrastructure component. Traditionally, we have considered telecommunications in terms of transmission and switching components. In the transmission components, one speaks of "copper wire," the traditional carrier of voice-grade communications; "microwave," or line-of-sight broadcast links that augment the wire network; "wave guides" or "coaxial cable" which provide for high-capacity transmission (the latter also serves for cable television), and "communications satellites," which are orbiting relay stations. Recently added to these are optical transmission systems such as line-of-sight light or laser links, and fiber optic transmission lines.

Switching, or management of the flow of messages in the network, is where "intelligence" enters into the system. The earliest switches were the well-known switchboards where wire and jacks were hand located to complete the circuit for calls. These were followed by electromechanical devices which allowed the caller, rather than a live switchboard operator, to connect calls by dialing a code. The kinds of decisions needed to patch two lines were built into the capabilities of the switch, and this evolved along electromechanical lines until the coming of all-electronic switches in the 1960s. Electronic switches are computer-like and capable of making decisions for call routing, tabulating tolls, and offering new services like abbreviated dialing. These are now giving way to entirely "digital" switches where even the message traffic is handled in a computer-coded rather than analog message form.

In the evolving intelligent network, switches increasingly enable:

- the integration of many different types of transmission technologies (broadcasting, wire, fiber optic) into a single "seamless" network

- the integration of many different types of telecommunications services (traditional telephone, cellular phone, local, long-distance, local area networks) into one, overall network

- simultaneous transmission of multiple messages on a single communications circuit ("multiplexing"), thus increasing the capacity of the network

- switching data or voice in a single network, supporting linkage of computers, databases, FAX machines, or imaging systems

- the conversion of messages from one system ("protocol") to another, as from a mainframe to a personal computer

- the determination of most efficient routing of messages in the network

- the system to reference, store, or communicate information on users, traffic, and tariffs, as, for example, in directory information, credit card checking, charging, and billing, selecting service providers (as in long-distance companies), or caller identification

- linking into computers and databases in order to offer information or transaction services (usually for a fee) such as news, travel schedules, shopping, statistics, weather, records, personal messages, or financial transactions, to name a few (sometimes called "videotex" services in general)

- performance of self-diagnostics to do error checking, self-correction, and to furnish the service provider with network diagnostic information

- the software capability for reconfiguring services, customer linkages, tariffs, or service options.

On the Planning Boards

Although there is no one master plan for continuing development of the intelligent network, most would agree on many of the following characteristics.

Increased Services Integration. As described above and illustrated in Figure 1.3, a major goal in network design is to be able to combine all types of systems and services into a single, overarching, network. Whereas different telecommunications systems proliferated over the years, the trend is for services to be combined in fewer, more powerful networks, including video services. Although this is currently possible on a limited basis (with compressed video, etc.), the larger goal might, for example, include all current cable-TV services in the switched network. This network will be both diverse in operation and of sufficient capacity so that bandwidth is not a limitation. Thus we could envisage high-definition television services on the coming network.

Multimedia Capability. With its increased integration and capacity, we can expect to have simultaneous availability of voice, text, data, video, and graphics. This should include availability of image processing (modification or enhancements), facsimile, and the ability to apply all such capabilities in computer-assisted-design (CAD), computer-assisted-manufacturing (CAM) or computer-assisted-engineering (CAE) operations.

Personalization. Increased personalization could include telephone numbers that easily "travel with you," or even a hierarchy of numbers, each devoted to specialized uses (business, social, family, emergency). This includes personalized uses of certain services such as a news file that is assembled to your profile of interests, stock market tracking and analysis, or easy access to selected CAD/CAM services.

Figure *1.3*. Integration of Network Services

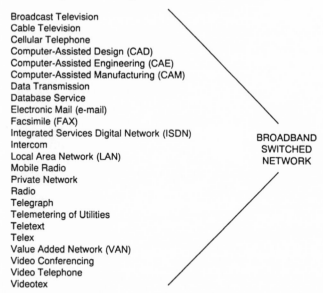

Broadcast Television
Cable Television
Cellular Telephone
Computer-Assisted Design (CAD)
Computer-Assisted Engineering (CAE)
Computer-Assisted Manufacturing (CAM)
Data Transmission
Database Service
Electronic Mail (e-mail)
Facsimile (FAX)
Integrated Services Digital Network (ISDN)
Intercom
Local Area Network (LAN)
Mobile Radio
Private Network
Radio
Telegraph
Telemetering of Utilities
Teletext
Telex
Value Added Network (VAN)
Video Conferencing
Video Telephone
Videotex

BROADBAND
SWITCHED
NETWORK

Portability. Access to the services should be widely available which means increased portability of terminals, services, and integration of broadcast and wired networks (like paging and voice telephone). Increased portability carries many locational or spatial implications, allowing for increased flexibility and decentralization. Achieving portability requires advances in terminal instruments capable of multimedia uses as well as independence from stationary power sources. One means for achieving portability is to have short-range digital radio links to voice or data terminals, even FAX machines. Increasingly this is called the development of a "Personal Communications Network" (or PCN), and it is on the planning boards of both regulated and unregulated telephone companies, the latter because they may be able to operate in local exchange areas.

Compatibility of Computer and Network Communications. Traditionally, this has been a matter of "protocol conversion" or creating an interface between one standard and another. Given progress in this area, the main challenge is to increase network capacity, speed, and error-checking standards to match those of computer networks, and approach the internal operating capabilities of the computers. In a broad sense, this makes the network take on the capability of an overall computer operating system, rather than these systems existing only within the computers themselves.

Improved Human Interface. Just as some computer users have found "mouse" and "pull down" interfaces an improvement over the entry of text commands, we can expect improvements in how we communicate with the network. "Hypermedia" techniques are a step in that direction, as is speech recognition and language interpretation (discussed below).

Speech Recognition. We can expect that the network will have an increasing capability to recognize speech, which, for practical purposes, means creating a textual version of utterances. Whereas creation (synthesis from text) of speech sounds is now commonplace in network use, it remains a challenge for speech to be recognized by computer technology. The problem is the great variation among speaker voice and dialect characteristics. Systems exist now that are sufficiently reliable for identifying the speaking of numbers that they can be put into practical use. However, systems that we expect to recognize a wide range of utterances still need a trial-and-error learning period and are prone to sufficient error that more research is needed before putting them into widespread use.

Language Interpretation. It is one step to transform spoken utterances into text but still another to have some type of "understanding" of that text so it can be acted upon. We should be able to anticipate some applications in this area to be incorporated in the future network. A simple version which one can find in desktop applications is where spoken commands are interpreted in operating a computer. But here the vocabulary is limited and often the syntax has to be agreed upon. These eventually can be accommodated by the network. Currently a priority is to develop a system where the network can answer simple spoken questions in a well-defined subject matter area, for example, directory information.

Language Translation. The Japanese have already designated language translation as one goal of Fifth Generation computer research and a likely service included in their national telecommunications network. While the Japanese have considered this in the context of speech recognition and synthesis modes, the service could also be a text-based one.

Networks as a Spatial Entity

Traditionally, the national network could be seen as a hierarchy of "hubs" and "trunks," which, of course, corresponds to cities and the long-distance links between them. Smaller markets are served by "spurs" from the hubs. Spatial patterning of the network reflected concentrations of population (local exchange) and patterns of communication and transportation that linked them (interexchange). The pre-divestiture organization of the AT&T network could be seen in a hierarchy ranging (Figure 1.4) from neighborhood switches up through a pyramid of switching tiers to the Class 1 switches at the top.

Figure *1.4.* Traditional Network Hierarchy

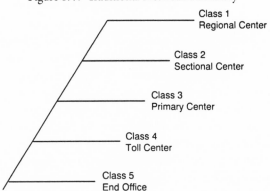

However, divestiture has altered the developmental pattern of the national network by opening the interexchange components of the network to competition while maintaining the monopoly status of the local exchange. This has changed which stakeholders (service providers, regulators, customers) may be most directly involved in applications of the growing network. Local exchange areas are now served mainly by the Bell operating companies ("BOCs") divested from AT&T and owned by regional holding companies ("RHC" or "RBOC"). Non-BOC exchanges are served either by large independents (e.g., GTE) or a wide variety of small independent companies and cooperatives. AT&T, along with its new competitors (e.g., US Sprint, MCI), serves the interexchange market.

As organized under regional holding companies, the local Bell companies are assigned to "local access and transport areas" (LATAs, for short) many of which are service areas defined around concentrations of population (similar to "metropolitan statistical areas"). Local exchange companies, considered to be operating within a "natural monopoly," provide service within LATAs (local exchange connections and access to the interexchange network). By contrast, AT&T and other long-distance providers haul the traffic between LATAs, considered an increasingly competitive part of the business. Within certain federal guidelines, local exchange companies continue to be regulated by state public utility commissions (PUCs), as are companies offering intra-state long distance. Interstate traffic is the province of the Federal Communications Commission (FCC).

An important generalization is that components of the network are divided among different business and regulatory interests. State PUCs are sometimes at odds with the Federal Communications Commission (and other relevant federal entities); local exchange and long-distance companies can have competing interests. For example, for years long-distance prices were used to keep the price of local service low, which was possible to do when the network was operated by one company. But in the attempt to introduce competition into the telephone business, the FCC has wished to move prices to the costs of providing individual services. State PUCs, often being especially sensitive to state and local politics, typically want to keep local rates low, so they resist changes. The same can be said of the higher prices of business services subsidizing residential rates, and urban services contributing to reducing the prices charged for rural services.

Much of the emphasis in studies of relations between telecommunications and economic development focuses on how governmental entities can sustain or promote economic development or delivery of public services at their respective policy-making levels. The national and interstate interests tend to reflect the activities of interexchange companies, whereas state and city interests often focus on local exchange providers, namely the Bell companies and independents. The regional Bell companies have been more or less of an "odd man out" because they have no natural overlay with governmental entities, and in many cases, the states their companies encompass may not house common interests.

A metaphor of the changing network was offered by engineer-attorney Peter Huber (1987), in his report on competition in national telecommunications. He likened new growth trends to the "geodesic" pattern of Buckminster Fuller's famous dome (Figure 1.5). This structure requires no

Figure *1.5*. Buckminster Fuller's Geodesic Structure

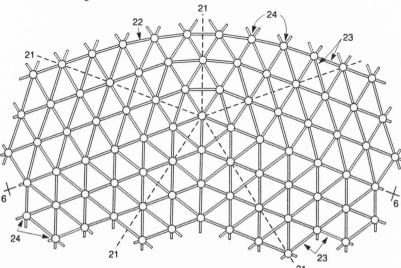

central frame; the analogy is a network that grows spatially in its hubs and interconnections rather than by expanding the switching hierarchy. Among the implications of this analogy is that "intelligence" is expanding away from the central part of the network and into hubs closer to concentrations of users.

Technological advances in telecommunications, many preceding divestiture, underlie the geodesic network. Although we have had noteworthy advances in the technologies of transmission, from multiplexing to fiber optics, it is switching that is now transforming the structure of the network. Microelectronics have simultaneously reduced the costs and revolutionized the power of switching and related components contributing to network intelligence. The consequence has been to increase switching alternatives and to move these alternatives out in the network closer to the telecommunications user. Thus we have a growth of numbers and sophistication of network hubs. Software-controlled, "open architecture," switches give us the capabilities to tailor and retailor network services to the user.

A geodesic network may be able to grow better than a hierarchical one in response to market forces, perhaps an advantage in our transition from monopoly to competition in U.S. telecommunications policy. Presumably, this network is flexible and "democratic," making it more responsive than the older hierarchical network to the rapidly growing and changing needs of users. This offers more locational options and flexibility for the delivery of advanced services. It offers the opportunity for multiple and competing service providers to advance growth, rather than depending upon planning by a single regulated company. It puts more power in the hands of telecommunications planners and regulators who are closest to the customers. Today, we are beginning to see states and cities paying more attention to the strategic uses of telecommunications.

Complementing the above trend are our observations that state and city officials are beginning to discover the importance of telecommunications infrastructure as a strategic investment rather than in the traditional "utility" view. This infrastructure is critical for developing new patterns of commerce and residential growth, and for delivering public services.

For example, in many large cities, telecommunications is among the bases for moving "back office" and light manufacturing out of the downtown areas and into adjacent areas where land is cheaper and workers more proximate. Whereas labor-intensive operations may be decentralized, the executive and headquarters functions may be concentrated in the central city area where it is important to have face-to-face contact with executives from other firms, bank, or government agencies. Jack Nillis (Nillis, Carlson, and Hanneman, 1976), who studied prospects for this shift in Los Angeles in the 1970s, now heads a "communications-transportation" trade-off program for the city. Mitchell Moss (1986) has written widely on the information-based transformation of downtown areas. And our own study of 12 North American cities illustrated how some are using telecommunications to distinguish themselves as a "gateway," "hub," or "port" (Schmandt et al., 1990).

This trend can also be seen for the nation as a whole in the NTIA Telecom 2000 report (NTIA, 1988), one of the first major federal documents to view national telecommunications as a strategic investment rather than mainly a utility.

THE COMPETITIVE CHALLENGE

As you can see from the introductory materials in this chapter, telecommunications has moved from being a sleepy utility into one of the major infrastructure components underlying our modern society. It is likely, too, that much of our advancement, including competitive positions with other nations in terms of economic development, human services, or national security, will be dependent upon our investments in telecommunications. In part, this marks our transition into the information age.

In the prior, largely agricultural then industrial ages, telecommunications received little attention beyond its identification as a utility not unlike electricity, water, power, gas, or the public roadways. Investment in telecommunications was treated as overhead by most organizations as a more or less necessary evil. Given adequate services, most organizations limited their concern with telecommunications to simply keeping the costs down. Public investments in telecommunications systems took on the nature of a utility either operated by the government, as in most countries, or by a regulated public corporation in the United States. Everyday citizens expected what the industry calls "POTS," or "plain old telephone service." But today we are witnessing a revolutionary change in the importance of telecommunications. As a consequence of the rapid advancements made in telecommunications and computing technologies, together with their coalescence and integration into modern businesses and service organizations, we are encouraged to regard telecommunications more as a strategic investment for progress.

We now face the challenge of making investments in telecommunications for specific purposes. These may range from the creation of new businesses—or the revitalization of traditional ones—to the undertaking of urban or rural, state or national, development. We look to telecommunications for the improvement of public services in health, education, and research. Such a challenge reflects the transformation of telecommunications from a taken-for-granted utility, like water or power, where the main concerns are low cost and availability, to telecommunications as an investment in competitiveness, productivity, economic development, or public service. It is an attitude whereby we expect a return on investment. Our success in this investment will be gauged not only in traditional terms of productivity, but in competitiveness in the ability of our cities and states, and entire nation, to compete globally for economic development and citizen services in the growing information age.

TWO

∎

Telecommunications as Strategic Investment

STRATEGIC (adj.)   necessary to or important in the initiation, conduct, and completion of a plan . . . of great importance within an integrated whole or for a planned effect.

TELECOMMUNICATIONS (n.)   communication at a distance (as by telephone or television) . . . a science that deals with communications.

—adapted from WEBSTER'S *NEW UNABRIDGED DICTIONARY*

In Part 2 we examine the types of returns to expect from telecommunications investments. We begin in Chapter 2 with discussion of facilitative functions of telecommunications in businesses and organizations. In Chapter 3 we expand this view to the role of telecommunications in economic development.

2

·

How Telecommunications
Adds Value

Scott Klopfenstein says right up front that he is no telecommunications specialist. His forte is making replacement blades for plastic recycling shredders, the main business of Prodeva, Inc., located in tiny Jackson Center (pop. 1,200), Ohio. But Scott is quick to add that Prodeva couldn't be competitive in this niche business if they did not make extensive use of 800 toll-free calling and their FAX machine. Their customers, mostly recycling companies, are highly specialized and widely dispersed, including some out of the country. Other than knowing how to manufacture a superior product, Scott explains, the key to staying on top of this business is close contact with customers as well as suppliers. "Giving customers an inbound 800 toll-free number as well as listing it in specialized mailings and ads has doubled our business over the last year," says Steve Bunke who heads Prodeva's one-man marketing operation. "Customers use this number for both ordering and trouble shooting," adds Steve, who notes further that if he doesn't hear from a favored customer at least once a month, he gives them a call on their outbound wide-area toll service. Scott explains that they use the wide-area dialing as well as FAX to coordinate operations with their suppliers. "Phone and FAX allow us to do business out of Jackson Center, where our overhead is about half of what it would be in a city."

Prodeva is a realistic example of how telecommunications investments can facilitate some specific aspect of a business or organization. In this chapter we describe this "value-adding" process. Chapters in Part 3 go on to give examples of telecommunications applications in business, science, residential, and public service contexts.

■ ■ ■

FACILITATIVE USES OF TELECOMMUNICATIONS

The Micro-Level View

If we take a micro-level view of how telecommunications adds value, we will typically identify particular points or processes in the operation of a business or organization that can be *facilitated* by the use of telecommunications. This is in contrast to a more macro-level view where in a community, region, state, or even nation, we attempt to identify the necessary availability of telecommunications as an infrastructure component for overall development. In the latter, we look at the aggregate effect of telecommunications uses or potential rather than in a specific business or organization.

The Prodeva micro-level example illustrated how telecommunications investment expanded the marketing operations of a small business. Other examples, by contrast, might be to use telecommunications to expand management operations among multiple retail store outlets, to use an on-line market information database to improve a company's business plan, or to use FAX as a link to suppliers, to name a few.

By contrast, on the macro-level—again the subject of the next chapter—would be examples of how the presence of a digital switch adds to a small town's attractiveness for business recruitment, how a city might encourage installation of a high-capacity fiber "ring" in its business downtown, how a state might upgrade its administrative network to facilitate small-town administration, how distance learning could enhance a state's public schools, or how 911 emergency services could increase a country population's sense of security and hence its attractiveness of new dwellers.

Value-Added Chain

One way to analyze opportunities for enhancement is to consider the operation of a business in terms of a "value-added chain" as illustrated in Figure 2.1 for a small manufacturing operation, such as was given in the opening anecdote to this chapter. The chain is a simple diagram of all of the interlinked steps in conducting the business. We could ask where improvements might be made in this chain so that we would have a return on that

Figure 2.1. Value-Added Chain

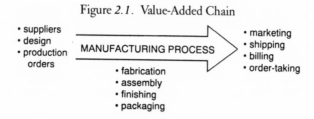

investment, or stated simply, an increase in productivity (ratio of costs to returns). The Prodeva example which opened this chapter could be identified as an investment to improve the marketing link in this chain.

For another example, we might invest in a small computer network that coordinates the placing of new manufacturing orders with on-line ordering of raw materials from key suppliers. Many such investments could produce more units relative to costs, speed up production, increase quality control, require fewer workers, or cut waste. Or we might invest in inbound 800 toll-free numbers for customers to report problems, thus gaining more customer satisfaction and information important for improving design or quality control. Or perhaps workers might attend training seminars over a video network to improve their production line skills. Again, as in the Prodeva example, productivity can be increased in supplier and customer links highly critical to this type of business. The returns are uninterrupted supplier operations, smaller inventories, and increased markets and sales. The bottom line is a more profitable business.

We can examine both strategies for and examples of telecommunications applications to increase productivity in a specific type of business or organization, an approach akin to microeconomic analysis. Obviously, some parts of a business are more susceptible to telecommunications-based productivity improvements than others, and some not at all. Also, because of differences in the composition of their value-added chains, some businesses have a more overall dependence on telecommunications (they are more "telecommunications-intensive") than others. We may also wish to study value adding in nonprofit organizations, as, for example, in how a public school might increase productivity by using distance education in the instructional process (as in the opening example of Chapter 8).

In this chapter and others (especially Chapters 4–8), we will see that in most instances it is not telecommunications itself that adds value. The improvement is in how telecommunications can *enhance* the performance of a worker, a production line, or an overall operation. We will also see that for the most part the costs of telecommunications are of less consequence than failure to obtain services or equipment either through lack of infrastructure, know-how, or breakdown. Costs of telecommunications are often less than 1 percent of the overhead of conducting a business, which can be trivial compared to wages, raw materials, or machinery costs. On the other hand, the absence of telecommunications can be very costly to a business, if not a reason for halting operations altogether. In the opening anecdote to the next chapter, we can see an example of how all business ground to a halt when the local switch failed. Finally, although many of the examples we examine are simple applications relative to the capabilities of the intelligent network, the basic rationale for added value is the same.

Information Technology Benefits

To visualize telecommunications benefits in other types of businesses or organizations, we can consider specific operations facilitated by telecommunications and relate them to a particular business. The concept of the value-added chain still applies, although we will not take the space to present different diagrams.

The basic function of traditional telecommunications is to overcome distance. This, in turn, may allow reduced costs of communication, time savings or compression, greater message frequency, greater selectivity, and, in general, increased connectivity in that more individuals, groups, or organizations can be in touch with one another. Depending upon the level of technology, telecommunications may greatly extend capacity for communications within units of time, as, for example, the amount of information in a televised image or burst of data as compared with a telephone conversation. Telecommunications may also support coordination either in real time, as in a telephone conversation, or in delayed or store-and-forward systems, as in electronic mail systems.

The intelligent network expands the foregoing capabilities in capacity, speed, simultaneous transmission of voice, data, and video, as well as bringing information processing services to the user (calculation, simulation, storage and retrieval, etc.).

What general types of benefits might we envisage, and thus consider for different types of businesses or organizations? Table 2.1 summarizes a set of examples drawn both from prior literature as well as the author's research experiences. Generally these involve applications for business expansion, decision assistance, improvements in shipping, storage, and routing, collaborative techniques, and the financial advantages of knowing market prices and being able to conduct on-line transactions.

Table *2.1.* Examples of Benefits in the Value-Added Chain

Business Decentralization

Telecommunications networks make it much easier to coordinate the components of decentralized businesses, branch stores, or intensive supplier links (as shown earlier in Figure 2.2). Management decentralization networks and methods allow a company to locate its sales outlets near lucrative markets, manufacturing near labor and suppliers, and headquarters in financial and design centers.

Market Expansion

Improved marketing by use of wide area telephone service, inbound 800 numbers, computer-assisted ordering, or telemarketing strategies may greatly expand the markets of a business. In rural studies (Chapter 11), we found this a key to the success of many small manufacturers.

Improved Decision Making

Most managers and planning personnel have more information at their fingertips and thus have the opportunity to improve the speed and quality of decision making. This can range all the way from information engineers might use in design operations, to gathering real-time information over a network that monitors information from the firm's production processes.

Cutting Distribution Costs

Computer-based scheduling and coordination of transportation units via mobile radio or cellular telephone connections can greatly improve the efficiency of transportation scheduling. Units can be rerouted, or a potentially empty unit on a return trip might be rescheduled to pick up new freight.

Cutting Inventory Costs

Using computer and telecommunications systems to keep track of inventory and to move it efficiently—a part of "just-in-time" manufacturing—has been one of the important changes in major manufacturing businesses. Even small business can get on-line services for restocking inventory and gaining analyses of slow-versus fast-moving inventory items, including productivity assessment in terms of shelf space.

Reduced or More Efficient Use of Labor

Often computing and telecommunications investments are either a substitute for labor or can make existing labor more efficient. For example, via local area networks the same production line worker can monitor more machines, or supervisors more worker-machine combinations. Better transportation scheduling saves labor costs of shipping personnel. Using telemarketing, a single sales person might multiply sales contacts tenfold.

Table *2.1.* Examples of Benefits in the Value-Added Chain (*cont'd*)

Increase Scope of Management

It is already well known that information technologies allow a manager to manage more operations or more people. Companies like AT&T have announced that through technology investments they have been able to cut entire levels of management from their organizations. The management hierarchy has flattened in recent years as there is less need for levels of middle management, and individual managers have increased their scope of responsibility.

Making Pricing More Competitive

Information networks often make it easier to calculate the true costs of manufacturing or wholesale products and thus provide a better basis for knowing price flexibility. Information from markets can indicate what the competition is doing. And changes in pricing can be disseminated almost instantly in many cases. Being able to be nimble in setting prices is a key means for gaining competitive advantage in almost any line of business.

Improved Opportunities for Training

In many advanced technology businesses, jobs or training to use new equipment cannot be learned entirely on the "job floor." Yet it is expensive to send workers away to training facilities. Telecommunications often makes it possible to bring training opportunities into the firm and directly to the worker. This can include video broadcasting, disks or tapes, telephone links, or computer-assisted instruction.

Lessen the Negative Consequences of Breakdowns or Stoppages

The ability to use toll-free 800 numbers for consultation, or to log on to information services, use FAX, or do remote diagnostics over the telephone improve the ability to get a production line started again, or to resume sales of a key product, or to straighten out shipping or inventory problems.

Benefits Reflected in Losses or Alternatives

Sometimes value can be placed upon telecommunications relative to the costs of alternatives or losses related to the absence of telecommunications services. For example, Kamal (pp. 64–66 in Hudson 1984), describes the value of telephone service relative to alternatives, including losses that might otherwise occur from the absence of telephone service (as in the opening anecdote to the next chapter). This includes alternatives summarized as follows:

- direct monetary benefits as reflected in differences between the cost of a phone call and alternative types of communications

- time savings that can be defined as a result of telephone use relative to number of working hours saved
- indirect monetary benefits gauged by the value of losses avoided if the telephone was used in an emergency
- indirect monetary benefits measured by more efficient use of capital or capital equipment because of the telephone.

TELECOMMUNICATIONS-INTENSIVE BUSINESSES

Expansion of Information Systems

As mentioned earlier, some businesses or organizations are much more prone to enhancement through telecommunications applications than others. One such type of business is where there is a need for coordination among decentralized units. This need has increased with the growth of conglomerates, and with national or international decentralization.

Figure 2.2 illustrates how telecommunications has increasingly supported decentralization over the past three decades. In the 1960s, technology investments were in individual support systmes like accounting, word processing, or manufacturing control. Mostly the inside units were "islands" of information processing. They were not telecommunications-linked. By the 1970s, we saw more interconnection of support systems on local area networks. That is, accounting could share information "on-line" with management or the word processing staff who might be preparing a fiscal report. This was the birth of management information systems that started as the integration of computer terminals into a network that provided access to separate word processing, data entry and retrieval, or electronic mail applications. Later, the terminal, now as a desktop computer, offered access to larger computer and telecommunications services, as well as local options such as word processing, spreadsheets, or computer-aided design.

Now, in the 1990s, we see the expansion of management information systems into networks extending far beyond the individual organization. The new systems, via networking and personal computers, also extend access to higher-level management. This results in enhanced control over data resources, cash flow, manufacturing operation, supplier or customer transaction, or market information, to name just a few instances. As we witness the globalization of industry, much of this is made possible by advanced telecommunications. For example, whether General Motors is manufacturing auto bumpers in Indiana or Matamoros, Mexico, the plants

Figure *2.2.* Expansion via Telecommunications

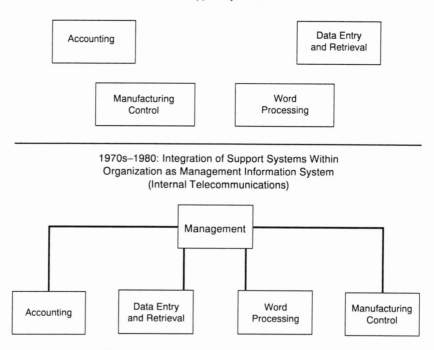

1960s–1970s: Independent Applications
To Support Systems

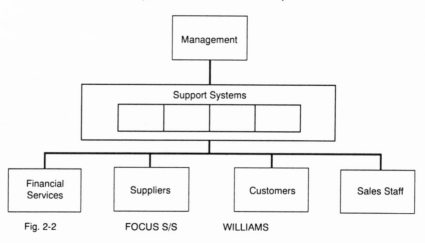

1970s–1980: Integration of Support Systems Within
Organization as Management Information System
(Internal Telecommunications)

1980s–1990s: Expansion of Management Information
Support System to External Environment
(External Telecommunications)

Fig. 2-2   FOCUS S/S   WILLIAMS

are integrated into the same highly efficient telecommunications network (Chapter 4). If they needed to do so, this integration could be in operation anywhere in the world, simply by linking into the global network. If facilities are not readily available, a satellite earth station can do the job. In today's rapidly expanding telecommunications world, we can expect to see intelligent networks strategically established to facilitate new dimensions of coordination and interaction.

Identification by Standard Industrial Classification Code

In some analyses of business uses of telecommunications it is useful to be able to identify telecommunications-intensive businesses in economic databases. Table 2.2 summarizes examples of telecommunications-intensive, service-oriented businesses in terms of Standard Industrial Classification (SIC) codes (as proposed in Schmandt, Williams, and Wilson, 1989).

Table *2.2.* SIC Codes of Telecommunications-Intensive Businesses

Information Services

Finance: SIC 60 (Banking), SIC 61 (Credit), SIC 62 (Commodities)
Insurance: SIC 63 (Insurance Carriers), SIC 64 (Insurance Agents)
Real Estate: SIC 65
Computer/Data Processing: SIC 737
Other Information Services: SIC 731 (Advertising), SIC 732 (Credit Reporting), SIC 81 (Legal Services), SIC 891 (Engineering and Architectural Services), SIC 893 (Accounting and Auditing)

Information Technology Equipment

SIC 3573 (Electronic Computing Equipment), SIC 361 (Electronic Distributing Equipment), SIC 365 (Radio and TV Receiving Equipment), SIC 366 (Communications Equipment), SIC 367 (Electronic Components and Accessories)

Research and Development

SIC 7391 (Research and Development Laboratories), SIC 7397 (Commercial Testing Laboratories), SIC 892 (Nonprofit Education and Scientific Research Agencies)

Media

SIC 27 (Printing and Publishing), SIC 48 (Communications), SIC 735 (News Syndicates), SIC 78 (Motion Pictures)

SOURCE: U.S. Bureau of the Census, Country Business Patterns, United States (No. 1) (Washington, D.C.: U.S. Government Printing Office, 1980).

Using the codes, we can draw summary statistics from census data on the growth of these business areas. And as discussed in several subsequent chapters, these are often the growth areas in the modern economy.

TELECOMMUNICATIONS COSTS AND BENEFITS

Return on Telecommunications Investment

If we expect organizations to invest in telecommunications with practical purposes in mind, it is necessary to conceive of how those purposes may be achieved. How does telecommunications add value? The answer is not simple, not because the processes are complex but because the effects of telecommunications investment must be seen in concert with a host of other factors. Again, telecommunications planners often say that rather than directly contributing value, telecommunications facilitates the development, manufacture, or sales of products, and the delivery of services. How, then, do we consider a return on investments we make in telecommunications? The literature varies, but generally modern answers to this question look beyond the simple, traditional "communications" function of telephone or data systems.

At the micro level, we can assess how telecommunications adds value through enhancement of specific aspects or operations of a business or organizational process. This, again, stresses the concept of information as strategic investment. The practical question is what is facilitated by telecommunications? Some answers are the "value-added" benefits as follows:

- enhance cost-effectiveness of an operation
- facilitate new forms of market access and distribution channels
- enhance competitiveness of a product or service
- enhance the overall competitiveness of a company
- facilitate the ability to adapt to change
- enhance the ability to introduce new products and services.

Although popular examples of applications often involve large businesses, the number and range of innovations are also found in many small business areas. New, small PBX systems, intelligent telephones, facsimile, and a variety of computer-communications services are available to the small business operator. We must remember that the information economy is not just huge companies and high tech; it is also "modest" tech contributions, even what one might call "low-tech" businesses (as in the Wal-Mart example in Chapter 4). New opportunities also extend to rural businesses, an important consideration for the economy. Among the applications to rural business are:

- control systems for resource conservation (e.g., irrigation systems)
- grower-processor-market transaction systems
- market research information
- weather forecasting and analysis
- remote management of decentralized facilities
- electronic funds management and transfer
- land value enhancement (e.g., installation of information, entertainment, and security systems for retirement communities).

Calculating Returns on Investment

In considering the facilitative applications of telecommunications, it is useful to think in terms of productivity, cost-benefit, or value-added perspectives where we attempt to measure facilitation. Several of these are described next so one can gain a sense of this perspective. (The examples are not meant as guides to calculation, as such studies will be more complex.)

The basic approach to measuring cost-benefit is the ratio of costs to financial return, and this is always seen relative to some type of alternative or standard. For example, suppose a telemarketing firm calculated that its average telephone charges per month were $5,000 for $100,000 worth of sales, a CB ratio of 0.05. Now consider the change if telephone costs were increased to $6,000 for speed dialing services, but sales increased to $150,000, resulting in an improved CB ratio of 0.04, or in simple terms the telephone cost of four cents for every dollar of sales generated. This reduction of one cent out of the original cost of five cents could also be characterized as a 20 percent improvement in cost-benefit.

Whereas cost-benefit is always in terms of financial figures, cost-effectiveness is cost relative to some other practical measure. Suppose, for example, that monthly telephone charges were $2,500 for 5,000 calls completed, or a CE ratio of 0.50. If speed dialing at $3,000 per month resulted in 6,500 calls completed, the CE would be an improved 0.46. Or the difference in four cents a call could be described as an 8 percent improvement on cost-effectiveness.

Cautions about Cost-Benefit Views

Despite the practical view offered by cost-benefit or value-added calculations, longtime information technology researcher Paul A. Strassmann

(1985) warns us that the ultimate impacts of technology investments may be far beyond the scope of these "near-term" types of calculations. For example, investment in a new telephone system might not show success in immediate cost-benefits or cost-effectiveness, but over several years open new markets for an organization. Increases in profits would not be found in near-term analyses and even in the longer run might be difficult to associate in such analyses with telecommunications investments.

Straussmann's general advice is that, in addition to cost-benefit types of analyses, we attempt to study the larger consequences which, although more subjective, may be a much more important indicator of the direction of a business than short-term productivity analyses. Referring again to the increasingly telecommunications-intensive environment of many modern organizations, enhanced "connectivity," improved coordination of company branches, or direct links with suppliers or customers may be more important in the long run than specific productivity ratios for a technology. This is another way of saying that as we attempt to gauge how telecommunications adds value, our ultimate answer may require as much imagination and thoughtful subjectivity as calculational approaches (unkindly called "bean-counting" by some).

THE HUMAN FACTOR

The Need to Manage Productivity and Reinvestment

Simply installing an information technology to facilitate some aspect of a business or organization, of course, does not automatically lead to a return on that investment. Typically, the increase in productivity has to be *managed* from the system, as do the returns or rewards of the implementation. The manager has to ensure that he or she achieves certain gains from the uses of technology, lest work expand to fit the newly available time, to use an old truism.

Moreover, if more products or operations are achieved per unit of time or cost, or more customers are reached, then there is the question of how that dividend will be used to advantage by the business or organization. This is the challenge of "reinvestment." For example, if a group of networked word processor operators can originate 10 percent more correspondence in a day, is that additional correspondence put to use in expanding the business? Is it needed? If not, then can the savings be realized in terms of reduced overhead such as that obtained by reducing the number of word processors or operators?

Information Workers and Managers

Peter Drucker (1989) has offered valuable advice on the changing nature of the workplace and its management as workers become more involved with information, or become specialists in it (as with programmers, computer or telecommunications specialists). As workers or specialists concentrate almost totally on information skills, they may consider themselves more as belonging to their information-based profession than to the business within which they work. They may be attracted to businesses where they can sense rewards for exercising their specialties. They may not be so tied to the overall business itself as much as they are to the opportunity for creativity and advancement in the kind of specialized information work that they do. As such, information workers know they are not dependent on one employer, and may anticipate multiple careers in their lifetimes. These workers often assume the direction of the company is up to top management, not particularly to specialists like themselves. The information worker bears a great deal of personal responsibility, because he or she must be the expert in deciding what information is relevant or not to the task. This person must also know how to get the information, verify it, and apply it to the problem at hand. It is not sufficient simply just to be an expert in the information; the information worker has to know how to put information to use.

Information-intensive organizations and their managers need a different type of leadership approach with information workers. Managers must understand the value of information workers and realize that they will be more independent and have less allegiance to a given organization. In many cases, the information workers, as specialists, will know more about their subject matter than those who manage them. They can take orders from superiors, but may expect more of an attitude of a "colleague" than a "superior-subordinate" relationship. In many cases, information workers will feel more rewarded for praise in their information skills and applications than in the overall productivity of whatever those skills are applied to. With more self-directed and educated information workers in an organization, there may be less need for managers and levels of management. Groups may work more as project teams and the manager may become more like a group leader who orchestrates the work of specialists.

There is still another way to look at the modern worker in the information age environment; this is the concept of "informating work," which is discussed next.

"INFORMATING WORK"

Another facet of business applications of telecommunications and information technologies is in the "informating" as against "automating" of
work (Zuboff, 1988). This is a topic that is likely to draw considerable
attention in the next few years.

"Informating" focuses on the change in the nature of work as it is
transformed from direct "hands-on" activities to the interposing of information technologies, so that the hands are on the technology which, in turn,
controls the work. "Automating," by contrast, removes the worker from
the picture. Using telecommunications to link computer-assisted production lines may serve to spread the environment of informated work, not
only in the specific production lines but in managing multiple lines.

One of Zuboff's graphic examples is the transformation of the worker
who can tell if a pulp mill "digester" (breaking down of wood chips into
pulp) is operating properly partly by its smell, a capability presumably
gained over long experience on the plant floor. What is the consequence
upon the nature of this work—and, of course, the worker—as this talent
becomes irrelevant, being replaced by an ability to monitor and control the
digester via an electronic interface?

It is not just a matter of learning how to perform with a computer
between you and your work; it is more like building skills with a particular
style of interface with the work. As we observed in site visits to two rural
Alabama pulp mills (Strover and Williams, 1990), in one sense it is like
being skilled at playing a computer game, but there will be further
differences between whether it is a fast-moving Nintendo-type challenge or
making decisions with a slower-moving (maybe "batch-like") interactive
"intellectual game." Put in simple terms, on one level, workers adapt to
entrance into the virtual world created by a computer interface; they are
neither working exactly with the reality of what their hands are touching,
nor is the computer simply an "electronic" hand on the machine. Their
hands, machine, and mind are interfacing in a "world" created by the
computer. Then there are differences in the parameters of what one might
be controlling, testing, or shaping in that world, and here is where skill
differentiation may again emerge. The finely honed skills of a pulping
operator may be visibly different from the finely honed skills of an operator
on the paper-making production line.

We still have much to learn about informating, but it seems inevitable
that with increased applications of computing and networking to produc-

tion lines and overall factory management, it is a topic to be pursued if we are to train the most effective workers and managers.

■   ■   ■

In all, we should remember that telecommunications adds value mainly by making people more effective. It is not just an investment in technology; it is one in enhancing human performance. Again, in the chapters of Part 3 we describe applications in business, science, the home, and in public service.

3

■

Telecommunications and Economic Development

On Mother's Day 1988, a short circuit ignited a fire that within minutes brought down a major telecommunications switch operated by Illinois Bell in Hinsdale, Illinois. Some 58,000 telephone circuits went dead— and soon so did commerce in that community just south of Chicago's O'Hare Airport and near the busy interchange of I-55 and I-294. Local service businesses were typically the first to shut down, succinctly described by a local realtor as "no phone, no work." Travel reservations stopped; local stores were unable to reorder supplies. Physicians and dentists could no longer shuffle their patient schedules. Parents tried to call schools and vice versa to the point where some parents took to the streets to retrieve their children. Probably most unnerving was that the residents in this suburban Illinois area could not pick up their phone to inquire as to what was the problem. Estimates were that businesses served by the Hinsdale central office lost approximately $2 million due to the switch outage, a very direct link between telecommunications and business. And this is not to mention the number of mothers who never received the flowers that families traditionally ordered by telephone.

Gauging the importance of the telecommunications infrastructure by its absence may not be the most comfortable approach, but it surely is an impressionable one. Such dramatic examples lead us to recognize the importance of telecommunications to modern businesses, emergency and public services, and residential life. In this chapter, we examine not the importance of the telecommunications infrastructure but how investments in it are a factor in promoting economic development. The chapters in Part 4 describe many specific applications in state, city, and rural environments.

■   ■   ■

ON INVESTMENTS IN INFRASTRUCTURE

Development as against Added Value or Productivity

One way to conceptualize the consequences of investments in telecommunications infrastructure for economic development is that this encourages an aggregation of all of the specific productivity, added value, or cost-benefit applications of telecommunications discussed in Chapter 2. The effect is to improve a business climate, industrial cooperation, technology transfer (taking ideas from the lab to market), or making an area more attractive to a work force or residential life. Economic development, in the large view, can mean the creation of jobs, higher-paying jobs, attracting or revitalizing businesses, or improving the quality of life, to name a few.

One common developmental application of telecommunications is for a town or area to promote installation of a digital switch, a fiber network, or both, so that the infrastructure is attractive to telecommunications-intensive businesses. Another is for the local telephone company to offer planning for telecommunications needs, to be encouraging about speed of installation and repair availability, and to offer continuing consultative services. An industrial park can contain state-of-the-art telecommunications services so that a new business can literally "plug in" and go to work. Another example is when a small town can offer toll-free calling (extended area service or "EAS" for short) to a nearby city. Often, however, these examples are more often seen in negative than positive forms, as when a town loses in a business recruitment competition because it does not have a digital switch or its telephone company takes little interest in the recruitment effort. Or a state may be unattractive to a business because its utility commission discourages extended area dialing service, intrastate tolls are high, or the regulatory environment is adversarial.

"Necessary but Not Sufficient Condition"

Do telecommunications infrastructure investments alone create economic development? The answer is generally no, because telecommunications is usually seen in a facilitative role; that is, its presence enhances

opportunities for payoff in other investments such as in business development, improving residential life, or efficiency in delivery of public services.
In our developmental research, the overall generalization has been that
*telecommunications is a necessary but not sufficient condition for economic
development*.

However, what is developed, when it is developed, who benefits, and
who pays for it remain topics of continuing debate. And so does the
definition of economic development which may variously refer to jobs,
attracting or retaining businesses, or enhancing the quality of life.

Players and Stakeholders

It is not at all unusual to observe economic development initiatives or
programs that pay little or no attention to telecommunications, especially
in their inception. Yet as many programs develop, sooner or later the
importance of telecommunications is recognized and it becomes a part of
the plan. This may change as U.S. telephone companies (mostly AT&T
and the regional Bell companies) become increasingly aggressive in promoting the importance of telecommunications for development. Also,
federal reports like the National Telecommunications and Information
Administration's *Telecom 2000* (NTIA, 1988) and the Office of Technology
Assessment's *Critical Connections* (OTA, 1990) advance a positive and promotional view of the value of telecommunications infrastructure for development.

Other players observed in developmental initiatives may include the
following:

Governmental Agencies. Some agencies of government, when faced with
telecommunications facilities planning (e.g., police and emergency
groups, state education agencies, human resource departments) may see
economic development as a goal, such as in a better-trained work force,
more attractive conditions for business locations, or increased productivity.

Educational Innovations in Telecommunications. These are increasingly
cited for economic development purposes, as in having the goal of "a
trained work force."

R&D Groups. There are a number of examples of advanced technology
centers where telecommunications is the research topic. Presumably, any
state or city that has on-site R&D services may encourage companies to
locate so as to take advantage of that environment.

Developmental Committees. Special committees sometimes appointed by state or city officials, or self-initiated, may develop plans for telecommunications and economic development, either directly or indirectly. These groups, when successful, may allow the careful consideration of longer-range alternatives in telecommunications planning, thus avoiding the short horizon of state regulatory proceedings.

Teleport Developers. As an analogy to seaports, there are positive examples of where special telecommunications receiving and sending facilities are located, sometimes with industrial space that offers particular advantages for telecommunications-intensive businesses or organizations.

Rural Initiatives

There has recently been increased attention given to uses of telecommunications to revitalize the rural economy of the United States. Rural sociologist Donald Dillman has written widely (1985; Dillman and Beck, 1988) on applications of telecommunications for rural development. Dillman sees the growth of the rural information infrastructure as important in such areas as:

- opening options for people to live in pleasant places distant from their traditional workplace
- providing market information for growers
- providing computer-based transaction (ordering, financial, database) services for rural businesses
- upgrading rural education, and
- allowing certain businesses to locate at a distance from cities.

It is perhaps ironic that most research literature on telecommunications and rural development emanates from third world studies of several decades ago, a topic well documented in Hudson's *When Telephones Reach the Village* (1984) or in the World Bank summary publication on the topic (Saunders, Warford, and Wellenius, 1983), all discussed later in this chapter.

GENERAL VIEWS OF TELECOMMUNICATIONS AND DEVELOPMENT

Telecommunications as a "Factor of Production"

If we turn to a more theoretical view of the role of telecommunications in development, we can consider it in economic terms as "a factor of production." A historical analogy, that of the railroads and their impact on economic growth, helps to illuminate the role telecommunications may play as one type of investment in infrastructure (Williams, 1988a). There is little dispute that the railroads were vital to American economic growth in the late nineteenth and early twentieth centuries. The production processes of that era required the development of a transportation infrastructure like that of the railroads. Economic growth at the time depended on growing markets and mass production techniques. To take advantage of economies of scale, corporations needed a way to link geographically dispersed demand. The railroads provided that link. Railroads, it should be emphasized, did not create demand, but merely enabled innovations to become economically viable by linking markets and thus permitting firms to realize economies of scale.

Telecommunications serves a similar function in today's economy. The importance of telecommunications, as with railroads, is linked to the nature of production processes. American manufacturing today relies much less on mass production than it once did. Much has been written about the movement of mass-produced goods overseas to countries with low-cost labor. Given the difficulty and undesirability of getting these industries back—American workers' wages would have to be reduced to the low level that was the incentive for the movement abroad—the hope for American industrial revival lies in high value-added, high-technology goods. The types of innovations that telecommunications can support are numerous: computer-aided design and manufacturing (CAD/CAM), education, work at home, and many more. Telecommunications is a crucial part of the infrastructure that will facilitate this revolution.

The intelligent network will facilitate the development of flexible production processes, seen by many as the key to American economic revival. Because America can no longer compete in mass-produced goods, it must compete in technologically evolving markets in which quality, rather than price, is more important. Producers must be able to respond quickly to changing markets. The time between design, production, and sales must be substantially reduced. Engineers and sales people must interact constantly to ensure responsiveness to the market. Quick and efficient communication is the key to making such a process function well.

Thus, telecommunications should be considered as an important factor of production. As industrial technology has developed, the production process has grown more complex and the organizational and informational tasks in production have grown more important. Informational activity has grown in size as well as importance. By 1980 one-half of all U.S. economic activity was accounted for by information creation and processing. This growth occurred primarily in the business sector. Just as the railroads were necessary to link the large markets that made mass production economically feasible, an advanced communications network is necessary to support the infrastructure for flexible production processes and segmented markets.

Telecommunications as Used to Attract or Retain Businesses

In a practical view, development often means efforts to retain or attract business to a city, state, region, or even a rural area. The recession of 1981–82 and the changing structure of the American economy caused states to become much more aggressive in economic development. With little prospect of recapturing the manufacturing jobs lost over the last two decades, many states and some cities (discussed in Part 4) have launched new initiatives to diversify their economies. These new initiatives have focused on strengthening the existing economic base, attracting new firms, encouraging the creation of new businesses, and providing assistance to distressed areas and workers.

Industrial recruitment is central to traditional economic development programs, although cities and states have developed a variety of new initiatives. Given that much of the new economic development activity was stimulated by the closing of manufacturing plants, successful industrial recruitment generates much goodwill for policy makers. Beyond being good politics, however, concern with manufacturing is sound economics. Even though the service sector accounted for 95 percent of new job creation from 1970 through 1984, service jobs exist, to a large extent, as a consequence of manufacturing jobs. Despite manufacturing's declining share of overall employment (about 21 million jobs or 20 percent of the work force in 1986), manufacturing generates another 40 to 60 million jobs. From one-half to two-thirds of these jobs are in the service sector. Most high-wage service jobs, such as those in finance, insurance, engineering, design, and communications, are linked directly to manufacturing.

The growth of producer services—defined as services, either internal or external to the firm, which aid in organizing the production process—is a significant post–World War II development in America's economic structure. In many cases, the service branch of a firm, like its manufacturing facilities, need not be located near corporate headquarters. This is now made possible, in part, by advances in telecommunications systems. States that are aware of a company's decision to expand its service branch can attempt to keep the service branch nearby. Alternatively, development advocates can try to lure service branches away from the corporate headquarters. One of the many inducements used to attract such firms is an advanced telecommunications network. In fact, Robert Ady, the CEO of P. H. Phantus, Inc., the nation's largest relocation firm, states that if the service branch or "back office" of a large firm is seeking relocation, the availability of digital telecommunications services is a "must" (FW personal communication).

Promotion of Scientific Collaboration and Technology Transfer

It is increasingly recognized that collaboration among scientists and their laboratories is important for modern research and development activities. Such collaboration can be enhanced by the existence of high-performance telecommunications networks (as discussed in Chapter 6), which can also promote the bringing of new inventions to market ("technology transfer"). This can be a type of telecommunications infrastructure investment as discussed in more detail in Chapter 6.

Benefits for the Work Force

We can also consider how telecommunications infrastructure developments can make an area attractive for the working population. Bellcore, the research branch of the regional Bell operating companies, in various pamphlets and presentations has suggested such benefits as:

- increases the capability for working at home
- makes available many remote services such as banking
- provides security and environmental control
- provides emergency medical communications
- makes it possible for a business to decentralize so as to make its branches more accessible to the work force
- provides entertainment on demand
- provides learning on demand
- provides connectivity to government information services
- makes available information and computing services in the home and small business
- facilitates coordination in two-wage-earner families.

Some Lessons from Developing Countries

Hudson (1984) makes the valuable point that many of the benefits of telecommunications go beyond individual users in a developing country to contribute to the society and the general economy. Here telecommunications is seen as facilitating delivery of social services, broadening the availability of market information to farmers, and bringing educational and training opportunities to remote areas ("distance education"). The benefits of telecommunications for developing countries have been explored in the earlier mentioned Saunders, Warford, and Wellenius (1983). For example, business persons can benefit from greater market information for buying and selling. Regional transportation planning and coordination can be enhanced by telecommunications services. Warning services can alert populations to potential disasters (tornadoes, earthquakes) or coordinate emergency services thereafter. Village liaison with regional or central governments can be made much more efficient. Reservation systems for local hotels can be brought into touch with modern travel agencies.

The CCITT (Consultative Committee for International Telephone and Telegraph), a committee of the International Telecommunications Union, has drawn generalizations (Hudson, 1984) on the national or regional benefits of telecommunications. Mostly, these are benefits from diminishing the cost of time and travel, and some overlap in part with those given earlier; for example:

- improving management, manufacturing, and government service delivery
- substitution or travel time and energy savings
- decentralization of business (distance management and information sharing)
- giving customers more information on products and allowing them to order goods
- expanding tourism through reservation services
- increasing efficiency in coverage for government administrative services
- impacting on agriculture production by improved ordering, coordination, requests for services or information, and availability of marketing information.

AN EXAMPLE OF A NATIONAL
INITIATIVE: JAPAN'S "INS" PLAN

INS = Information Network System

Telecommunications and development can reach national proportions as exemplified in Japan's "Information Network System" master plan. The Japanese intend that by early in the twenty-first century they will have linked their major and regional cities with a state-of-the-art national fiber optic telecommunications network. This Information Network System, combined with communications satellites and other new broadcast technologies, is scheduled to replace the existing network built in the years immediately following World War II.

The INS plan, initially introduced in 1981 by the Nippon Telegraph and Telephone Company, will incorporate all the leading-edge information technologies, including an all-digital architecture, supercomputers as well as Fifth Generation computer integration, simultaneous video, data, and text signals systems, and artificial intelligence. The cost of this system is envisaged as ranging up to $150 billion, yet is expected to yield up to $300 billion market in a new era of telecommunications equipment, services, and value added to traditional businesses.

The Japanese Technopolis Plan

The INS is at the heart of the "Technopolis" infrastructure advanced by the famed Japanese MITI organization—the Ministry of International Trade and Industry. This twenty-first century national objective is MITI's aggressive undertaking to create 19 "Silicon Valleys" across the Japanese archipelago (see Figure 3.1). These regional high-tech industrial centers will be linked by an advanced technology telecommunications trunk, and each of the regions will draw as well as share its information services from a vast grid of advanced telecommunications technologies (Tatsuno, 1986).

The INS is designed to achieve several strategic goals. One is to offer a readily available implementation market for information technologies advancements coming from Japanese laboratories. Just as the Fifth Generation computer (stressing artificial intelligence, parallel processing, voice input) is a goal in itself, the implementation of those computer innovations into scientific, industrial, and public service settings is a second-stage goal. The technopolis concept is a highly advantageous environment for achieving this goal, as it gives the Japanese early experience in the applied capabilities of its technologies, a basis for technology refinement, and the type of information highly valuable for international marketing. (One often sees English logos on advanced Japanese technology, surely a sign of their marketing plans!)

A second strategic objective of the Information Network System is to bring the most advanced communications and computing services to the Japanese population. This includes not only services for industries, but for Japanese homes as well. The twenty-first century Japanese home should

Figure *3.1*. Japan's "Technopolis" Network

have available a wide variety of new information technology services, for example:

electronic security monitoring

video telephone

facsimile communications

videotext services

telemetering of water, gas, and electricity

home shopping

home banking

computer-aided instruction

teleconferencing

entertainment and cultural television services

computer-aided instruction.

There is no end to the services the Japanese envisage for this new network. Perhaps most venturesome is to take the voice recognition and synthesis capabilities as well as the language transition capabilities targeted for Fifth Generation computers and to add them as a "language translation service" on the voice network. Conceivably, one could pick up the phone and in at least a limited vocabulary carry on a translated English-Japanese, Japanese-Russian, or German-Japanese voice conversation.

The technopolis concept will itself depend on the success of the Information Network System. Any computing capability located in the 19 technopolis centers—or for that matter in Tokyo—will be immediately available throughout the network. Unlike our current problems with computer telecommunications, there will not be barriers of interfacing one brand of computer with another, patching together ad hoc networks, or having different command and control systems. The "intelligent" advanced telecommunications network will automatically accommodate all such conversions.

Together, strategic telecommunications and the technopolis plan are a key component in Japan's "take lead" strategy for positioning itself in the twenty-first century. The return on investment is a new competitiveness for their businesses, a wide variety of social services for their population, and an overall catalyst for national development.

U.S. POLICYMAKING AND ECONOMIC DEVELOPMENT

What Do Planners Mean by "Economic Development"?

Given the growing recognition of the importance of telecommunications to the economy, the question is posed as to how this affects policy making. In the recent decade, two issues have stood out in U.S. activities. The first has been the need to specify what is meant by economic development. Because the term has become so politically popular in recent years, it has come to mean many things to many people. New jobs, increased gross product, technological innovations, higher wages, and an improved quality of life all could come under the rubric of economic development. Differentiation within these categories is also possible. For example, a region may target specific types of jobs, such as those found in the producer services sector or those offering high wages. Also, the telecommunications industry itself can be considered a focus for development. These differences elude an overall definition of economic development acceptable to all concerned. On the other hand, one can examine development simply in terms of what the proponents claim as their practical goals.

Who Will Take the Lead?

The Federal Government. A mainly developmental perspective on telecommunications has not been a federal priority. Although there are various reports, such as the National Telecommunications and Information Agency's *Telecom 2000* (NTIA, 1988), that describe opportunities for strategic investments in telecommunications for national development, there have not been major initiatives to put these ideas into action. Mostly, activity at the federal level has centered around the divestiture of AT&T and the follow-up actions of the Federal Communications Commission (charged with national and interstate policy), the Department of Justice (examining antitrust issues, the legal background to divestiture), and the court of District Judge Harold Greene (reviewing compliance with the antitrust divestiture agreement or "Modification of Final Judgment"). The lack of federal direction in the U.S. contrasts with the aggressive planning of the Japanese as well as the European Common Market.

State Regulators. Members of regulatory commissions have not traditionally had economic development in their mandates. In most states, when a link has been posed between regulation and development, it has come from outside the commissions. Mostly, it has come from the industry and their supporters favoring regulatory flexibility. On the other hand, despite having a historically narrow view of regulation, public utility commissions are possible sources of new developmental initiatives within state governments. They may be directed by the executive or the legislature to implement or study innovative policies. Moreover, commissioners themselves may favor policies linking economic development with telecommunications. In some states, commissioners are elected officials and subject to the same political pressures as legislators. In other states, commissioners are appointed by the governor, perhaps subject to legislative confirmation. In any case, economic development could become a component in regulatory activities.

State Development Agencies. Government agencies are another group to consider. They have expertise and a public mandate. However, state economic development agencies have historically focused on manufacturing and public-financed infrastructure development, such as roads, bridges, and sewers. The techniques that bring these development efforts to fruition may not be appropriate to telecommunications. But currently, as the strategies for state economic development are being cast in broader terms (including policies for technological innovation and human resource development), telecommunications is seen as an increasingly important part of the state infrastructure.

State Legislatures. State legislatures may have standing committees or subcommittees on telecommunications or utilities in general. Moreover, since legislatures are sensitive to political pressure, lobbying by the telecommunications industry, large telecommunications users, or other constituents can lead to legislation promoting economic development via telecommunications.

City Planners. U.S. city governments have typically played a benign role in telecommunications development, save to become involved when it is necessary to rule on rights-of-way, cable-TV franchises, and the city's own needs in telecommunications. Recently, however, there has been an awakening to the importance of modern telecommunications for attracting new businesses, retaining old ones, or for improved delivery of citizen services. This is the topic of Chapter 10.

Special Task Forces. A final group to consider is special task forces, namely individuals brought together to consider telecommunications initiatives related to development. The "Intelligent Network Task Force," assembled in California, is an example of a citizen group, in this case prompted by a request from Pacific Bell (discussed in Chapter 12).

■   ■   ■

In all, the uses of telecommunications for economic development will only come when we have an awareness of its value among the above players, and when they are willing and able to coordinate their efforts in making the investment. Again, examples of investment in state, city, and rural areas are the topics of chapters in Part 4.

THREE

■

Telecommunications Applications

The technological capabilities of the new telecommunications are but part of the picture. More important is what we do with these new capabilities. We can already see changes in the ways we do business and in how science brings new ideas to the marketplace ("technology transfer"). Small businesses are in a state of change as modern management methods are made available by network services. Perhaps the largest applications market by far is the home, where the new telecommunications offers expanding opportunities for information, transactions, and metering services. Finally, just as information technologies can bring new productivity to business organizations, they can do the same in not-for-profit applications, or the public service sector. This applies not only to traditional public service organizations in governance, health, or the environment, but especially in education. Perhaps on no front are we challenged as in educating citizens for the information age.

4

■

Networking Large Businesses

The plant manager proudly proclaims that if a red Pontiac Firebird is sold in Los Angeles today, in less than a week he'll know that his GM plant in Matamoros, Mexico, will have to turn out a red bumper. It's called "just-in-time" manufacturing, he explains, and it is made possible only by a sophisticated international network that links all GM facilities. The designer of this network is Electronic Data Systems, or "EDS" for short, originally the brainchild of computer genius and Texas billionaire, Ross Perot, who sold his company to GM, then departed over differences with GM management. The manager goes on to explain how it doesn't matter whether a GM plant is in Detroit, Singapore, or Matamoros because the network makes "distance irrelevant." In fact, much more than manufacturing orders are exchanged between the manager and GM headquarters. The network supports all their personnel and financial records keeping, new design information, troubleshooting, and even training. About all it cannot do, jokes the manager, is change the dollars for pesos that the Mexican government requires to pay the workers.

Business provides what is probably the most diverse environment for telecommunications applications, from communication and control functions, to the furnishing of information services. Large businesses, including those doing business on a global scale, have been the most fortunate recipients of the new telecommunications.

■ ■ ■

CHANGING THE WAYS OF DOING BUSINESS

The new telecommunications is both changing our ways of doing business and the businesses that we do. You may remember in Chapter 2 our description and illustration (Figure 2.2) of how network services can be used to manage the decentralized operations of a large business. Our opening example of the Mexican GM factory is an illustration of such decentralization and its importance to the global expansion of U.S. industry. We can also see a similar application (Figure 4.1) as the "hub" concept of managing a business. This might be the managerial structure of a large decentralized business, like GM, or it could be the operational structure of branch banks, or a large retail chain like Wal-Mart (which we examine shortly). The point is that telecommunications has given us many new options for organizing and managing our businesses. Some of the current and coming services that support large business are listed in Table 4.1.

Large businesses are able to coordinate their many departments and locations, making it increasingly possible to manage an organization whose spatial dispersion reaches global proportions. Sales outlets can be located near markets; manufacturing near work forces and suppliers; and headquarters near financial and design centers. Branch offices or chain-store outlets can avail themselves of modern services via network applications. Telecommunications increasingly makes it possible to locate components of business in low-cost rural areas, as described in Chapter 11.

Figure *4.1.* "Hub" Organizations

Table *4.1.* Examples of Network Services for Businesses

Typical Services

—Voice and data virtual private networks for large company management
—Local or national "point-of-sale" transaction networks
—Telemarketing using wide area telephone service
—Toll-free (800) calling for customer inquiries, orders, or trouble-shooting
—On-line accounting or personnel management services
—Teleconferencing among business offices
—Network services for employee training
—Network tracking of shipping and handling
—Use of scientific high-performance networks to connect research laboratories with manufacturers ("technology transfer")

Coming Advanced Services

—Virtual broadband, multimedia private networks for large company management
—Increased portability of services as with cellular FAX, PCs, and point-of-sale transaction terminals
—Integrated portability as personal communication networks
—World-wide paging and radio-telephone services
—Easily available, "dial-up," video conferencing
—Advanced multimedia network linking for computer-assisted design, computer-assisted manufacturing, and computer-assisted engineering
—Satellite-based transportation tracking
—Multimedia catalog selling and shopping
—Access to parallel and supercomputing services

Most large corporations now use network services to coordinate management, accounting, personnel records, inventory control, transportation and shipping, and training. As described in Chapter 2, advanced telecommunications helps a manager to manage more people, processes, and systems. Uses of the network have allowed major companies to cut entire levels of middle management because of new managerial efficiencies and expansion of scope. Via the network, management can be expanded beyond the spatial limits of the immediate office or plant to include, for example, branch plants, job sites, or employees working at home. It is becoming customary in some large, decentralized businesses to have regular management meetings via full video teleconferences, thus cutting costs of time and transportation for traditional meetings. Managers now have more information directly available to them over the network, more "decision assistance" from individuals and intelligent systems, and can communicate more efficiently with larger numbers of associates. Many organizations with "back office" paperwork units (e.g., banks, credit card processing, insurance claims) have sought separate locations for these units where space and labor costs savings can be gained. In most such cases, access to advanced telecommunications services is a necessary requirement for relocation. And relocation need not be in the same city; it might be in the countryside.

Much global competition among advanced industrial countries seems to be focusing upon capabilities for "flexible manufacturing," that is, the ability to move rapidly from one set of specifications to another in producing limited amounts of goods. This capability is often supported by manufacturing technology that can be easily "reprogrammed" to meet different specifications or accomplish varying tasks. Coordination and management of flexible manufacturing, especially in a large decentralized company, and in tandem with computer-aided design, requires advanced telecommunications services.

Design activities can now be more decentralized in many organizations. This has been made possible by linking collaborating engineers or scientists over the network. Some current computer-assisted design (CAD) systems can be operated over a network where, say, designers can share graphics, calculations, photographs, or run programs as they collaborate in their work.

Network services are increasingly available to assist in selection and management of the most efficient routes for product distribution, warehousing, and special shipping needs. Cellular telephones are used to keep contact, route or reroute trucks or salespersons in the road. There are new satellite tracking technologies that report the location of transportation units, estimate times to arrival, or calculate fuel and time efficiencies. Part of the acknowledged success of the Wal-Mart discount chain—to be discussed in detail later in this chapter—is due to highly efficient movement of inventory from suppliers via warehouses to the retail outlets, all managed over a private satellite network.

Next are some "snapshots" to illustrate the flavor of large business applications.

SOME LARGE-SCALE SNAPSHOTS

Lucky Goldstar: Look Out for Korea

In the golden twin tower closest to the Han River in Seoul, Korea, telecommunications and computer engineers have been fashioning a global telecommunications network that will allow the giant Korean conglomerate, Lucky Goldstar, to put a customer or supplier on-line nearly anywhere on earth. "It's like the American Hospital Supply" legend, says an executive from GM's Electronic Data Systems, which is partnering with Lucky Goldstar to transfer EDS network know-how to Seoul. "Once we provide a customer with the convenience of a direct telecommunications data link, they'll think twice before going elsewhere. We will just be all too convenient, and, of course, Lucky Goldstar is one of the most competitive companies on earth." "We can transfer orders, inventory tracking, funds, and coordinate CAD/CAM operations over the network, even voice when necessary," he adds (personal visit, September 1988).

An important twist to Lucky Goldstar's approach, as compared with typical EDS customers or partners, is that the network managers want to be able to operate in any environment, even one with a 1950s telecommunications infrastructure, like the People's Republic of China. Thus they are designing a network and software that can adapt to nearly any stage in managerial uses of telecommunications. As clients progress in telecommunications and computer sophistication, the network can mature with them, if not help the developmental process.

The other goal of Lucky Goldstar's telecom group is to sell their know-how and services to other global companies, called telecom "outsourcing" in this business. They will seek multimillion dollar contracts over five- and ten-year spans so as to be a "turn-key" provider of telecommunications and data services. EDS executives in Seoul see this as up to a $15 billion business in the late 1990s. And it is not a coincidence that Lucky Goldstar also manufactures many of the relevant equipment components.

Bechtel: Global Engineering

For information-intensive companies that employ field personnel, telecommunications can improve service for both the firm and its clients. In California, Bechtel Group engineers are linked to company computers by satellites and other lines when they are at remote international project sites. The effect is that this large firm with diverse operations centralizes some investments while decentralizing their use. Engineering, including CAD/CAM operations, can be done over the network at most company locations.

Bechtel was an early potential customer for ISDN but was not able to get it via the local network. They partnered with AT&T to develop their own ISDN in order to link computer-aided design and engineering workstations, even though they claim to have preferred the public network (Schmandt, Williams, and Wilson, 1989).

Weyerhaeuser: Revitalization of a Traditional Industry

In Washington, the name Weyerhaeuser is synonymous with lumber and wood products, yet the company's business extends far beyond those products to include housing, mortgages, nurseries, and shipping. Mike Knight, a lobbyist for Weyerhaeuser, says that when the company chooses a site, the telecommunications facilities are a concern, but often trade-offs are made if some of the other criteria—skilled work force and low-cost land, for example—are met. Occasionally, the firm even acts as a telephone company when it locates in an area with no current service. Additionally, since 1979 Weyerhaeuser has been providing minicomputer access so customers of pulp and paper products may order without having to go through a salesperson. These computer hook-ups require a level of service not found in all of rural Washington. Consequently, Weyerhaeuser supports Washington's move toward improving the infrastructure of the telecommunications system throughout the state (Schmandt, Williams, and Wilson, 1989).

Chase Manhattan: Going Global

Not only can networks directly serve the management of large corporations, but they can provide the financial services as well, including international market trading. For example, among New York's financial institutions, Chase Manhattan Corporation has in recent years become one of the world's leading financial service organizations. With assets of $95 billion and a presence in more than 70 countries around the world, Chase has engaged in a business strategy that seeks to integrate the latest information technologies with its basic products and services. Tying the whole operation together is Chase's international communications network. Included in that network is MCI's first international T-1 service between the United States and the United Kingdom. Chase uses its T-1 service mainly for voice transmission, but the service can be used for any number of applications, including data transmission at various speeds, facsimile transmission, and video conferencing. Richard A. Pandullo, a Chase Manhattan vice president, is continually looking for ways to improve the bank's network, always keeping in mind that his company's business strategy must drive the technology, and not the reverse. "It's not where you stand today, it's the direction you're moving that counts," says Pandullo. "That's how you keep your edge" (Schmandt, Williams, and Wilson, 1989).

Chicago Mercantile Exchange: Midnight Trading

Telecommunications also plays a crucial role in increasingly global financial markets. The Chicago Mercantile Exchange has established communications links between Chicago and Asian trading centers so that Chicago traders can have up-to-the-minute information on the Far East market. Advanced communications technology enables Chicago Mercantile to implement a midnight shift of traders keeping abreast of Asian financial markets. Another interesting feature of telecommunications usage at the Mercantile is the design of the network. Although it may be more efficient for all the trading houses to share one private network, houses have instead constructed a series of private networks for proprietary reasons. Traders prefer the security of their own networks to the potential cost savings of a shared network, showing again how businesses are increasingly viewing telecommunications as a strategic part of their operations.

James River Corporation: Rural High Tech

The Naheola mill of the James River Corporation is the largest employer in Marengo and Choctaw counties of Alabama. The company, headquartered in Virginia, has been the second largest paper company in the world. Paper mills like Naheola are also increasingly a technology-intensive industry, including both chemical processing and uses of information technologies. Computer interfaces are involved in many of the production processes, including pulping, papermaking, and "converting" (the latter refers to cutting and partial pasting for container making). The interfaces are partly linked on a local area network. The various mills are, in turn, linked to company headquarters. In some respects, the management network extends from headquarters all the way down to production processes. The Naheola mill represents what economist Amy Glasmeier (1991) classifies as a rural "high-tech" industry, and computers and networking reflect the "informating" work concept (Zuboff, 1988) discussed in Chapter 2.

HIGH TECH FOR LOW TECH:
DOING BUSINESS OUT OF BENTONVILLE

Networking Wal-Mart

Not all applications of modern telecommunications need to be in high-tech or global industries. One of the greatest success stories is in the management of Wal-Mart,[1] the discounter that has changed the retailing patterns of rural America. Located in a small town in northwest Arkansas, west of the Ozark mountains, Wal-Mart Stores, Inc., quietly runs a retail empire. The Bentonville-based general merchandise discounter is a familiar name to most people, particularly in the U.S. Midwest and South where the discount store is a familiar sight.

Since the first Wal-Mart discount store opened in 1962 in Rogers, Arkansas, the company has been on a fast track. The company has climbed from being a relatively unknown regional chain, operating 276 stores in 11 states in 1979, to the nation's number three mass merchandise retailer, even though it currently only has stores in about half of the continental United States. By 1990, Wal-Mart, then the most profitable retailer in the United States, was expected to become the sales leader as it was due to surpass Sears, Roebuck & Company and the K-Mart Corporation in merchandise sales volume. At that time, Wal-Mart announced plans to enter California and the East Coast, setting the stage to become a true national merchant, and promised to open more stores in 1990 than in any previous year.

The 27-year-old retailer's spectacular growth and expansion is a remarkable story that has drawn national recognition. Many would agree that one of the most important factors in Wal-Mart's success has been its ability to bring its expert management methods to rural areas by use of a national private telecommunications network. Wal-Mart's discount stores are typically located in small rural towns, although more outlets are being opened in and around metropolitan areas. The rural setting has advantages such as lower operating costs, the loyalty of a committed work force, and better environmental amenities. But what is more significant is that the dispersed and remote environment perpetuates the need for Wal-Mart to better integrate modern communications technologies. Wal-Mart is one of the first chains to recognize the potential of small towns and has found enormous profit and growth opportunities in the out-of-the-way market that others had ignored. The company has used advanced telecommunications technologies to turn traditional rural barriers into advantages.

Wal-Mart has been in the vanguard of innovative technology use, especially in the 1980s. It was the first major retailer to adopt the UPC (Universal Product Code) bar code scanning system which not only speeds up checkout but allows the company to track merchandise. Its extensive use of computers in its headquarters has facilitated fast internal communication. Yet the centerpiece of the company's technology is its VSAT (Very Small Aperture Terminal) satellite network. VSAT is a business satellite network that features a central hub facility with a terminal at each branch outlet.

In 1984, Wal-Mart took a big step in launching its VSAT system for data and voice communication. The initial trial turned out to reduce data and voice costs significantly. The company then capitalized on its investment by searching out and developing several other benefits that could accrue from using the technology. By 1990 Wal-Mart's VSAT network supported two-way data and voice communications between its headquarters in Bentonville, Arkansas, and 1,600 Wal-Mart stores, Sam's Wholesale Clubs, Hypermarts, and distribution centers in 25 states.

Notes on VSAT

Early applications of satellite communication were mainly restricted to government, military, or space science use because they required large and expensive earth stations. As the technology improved, the size and the price of the dish dropped. The first VSAT earth station with a two-foot dish for data reception was introduced in 1981. The existence of the very small aperture earth terminal made it cost-efficient to bring satellite communication directly to the end user's location and offered American corporations an alternative to leased-line communications.

The nature of VSAT technology requires a controlling master station to communicate with thousands of remote stations. The capabilities and operation of a master station create the network. VSAT communication supports data, video, and for Wal-Mart, voice communications. In the early 1980s, the primary users of VSAT were the financial, news, and media industries for two-way data and one-way video transmission. Two-way voice and video are also possible at the expense of higher earth station and satellite capacity cost. Standard video requires a full satellite transponder in terms of both bandwidth and power, with a significant increase on earth station cost. Voice applications are almost exclusively limited to "closed network" requirements, where a private network is advantagous from a cost or availability standpoint. Wal-Mart's voice traffic, by design, is limited to supporting communications from the corporate headquarters to the branch locations.

Corporations adopt VSAT technology for specific applications relevant to their competitiveness and growth. The technology that was initially implemented in the finance and banking industries has expanded to support two-way applications in retail, hotel, and data processing industries. Today, many large businesses install VSAT networks as the main means for bypassing the public telephone network. This gives them advantages in costs as well as control over their communications facilities. The FCC's deregulation of Ku-band satellite frequency played an important role in boosting the development and use of VSAT systems.

More on Applications

The use of Wal-Mart's VSAT network is probably the most unique and cost-effective application. It is unique because to date Wal-Mart is the only corporation that uses VSAT for telephone purposes on such a large scale. This application has also proved to be extremely cost-effective. Since the satellite is not distance-sensitive, it is ideal for Wal-Mart's dispersed multi-location operations. Also the costs of satellite transponders are fixed and not affected by the volume and length of phone calls. Wal-Mart estimates that the VSAT network saves two-thirds of its long-distance telephone costs. Further, it allows Wal-Mart more control over its communications. The company no longer has to rely on the 400 telephone companies that it used to deal with for voice communications. However, Wal-Mart still maintains links with the public switched network which it uses for calling to and from customers, and as a backup system for its VSAT network.

The voice application of VSAT is not yet widely used by other companies. There are at least two reasons for the lack of interest among VSAT users for this specific application. First, voice requires substantial bandwidth as well as sophisticated and expensive equipment. Second, there is a slight delay associated with a satellite-transmitted telephone call. Many corporations have decided not to pursue this application for these reasons. However, Wal-Mart is able to use the satellite bandwidth in such an efficient way that both data and voice can share the same satellite capacity. Furthermore, Wal-Mart uses VSAT only for internal communications, namely, between headquarters, distribution centers, and branch outlets. Since Wal-Mart owns all its retail outlets, the company can require its employees to use VSAT for voice communications despite the undesirable delay.

The private network simplifies the dialing procedure for any extension within the system. An employee at headquarters simply can directly dial, with a six-digit number, any Wal-Mart store. If employees from a store need to call headquarters, they only have to dial a four-digit extension number to reach a certain department or executive in the General Office. The VSAT phone system in each store is programmed to reach only the Arkansas headquarters.

In the data network, the previously leased terrestrial network has been completely replaced by the satellite system. VSAT provides a much faster, more reliable, and higher-capacity two-way data transmission with the formerly leased land lines. The link between Wal-Mart's high-capacity computers and the VSAT network keeps headquarters informed of all transactions in each store every minute of the day or night. Unlike the previously used leased-line network, the VSAT network also allows stores to place orders simultaneously, speeding up the ordering process.

Video is a more recent form of communications supported in Wal-Mart's VSAT network, and it is used to provide video training, teleconference and marketing applications, and to deliver executive presentations. Each Wal-Mart store, Sam's Wholesale Club, Hypermart, and distribution center has been able to receive twice-a-week video motivational or instructional messages from Wal-Mart's former chairman Sam Walton, Chief Executive Officer David Glass, or other executives. Although the VSAT network has the capability to transmit two-way video, so far Wal-Mart uses it only for one-way delivery.

Wal-Mart has already realized a return on investment through lower communications costs in voice and data. It was even able to capitalize on its investment by searching out and developing other benefits that can accrue from using that technology. In 1988, it added to its satellite system a credit authorization system, which is referred to as "point-of-sale" in the industry. In this application, it takes five seconds on average to verify a card while previously a phone call to do the same could take up to a few minutes.

The satellite network also has been used to test a truck location system whereby each delivery truck's exact location would be relayed to headquarters every 15 minutes. Although Wal-Mart did not pursue this application, VSAT does have the ability to provide a truck-tracking system. The next application that will piggyback on the VSAT network is a Muzak system, which was to appear at the time of this writing. When installed, all Wal-Mart shoppers will be listening to the same music no matter in which of the 1,600 stores they are shopping.

Competitive Advantages

According to Jay Allen, the former manager of Wal-Mart's satellite department, there are four major advantages of the VSAT system. First, it is flexible. The system allows easy and quick addition of sites. Also one can change band rate with a key stroke and add new applications such as a bandwidth or channels without worrying about overtaxing the system. Secondly, it is cost effective. Allen estimates if Wal-Mart tried to do what it does now with the satellite over phone lines, there would be a 300 percent increase in cost. He estimates that on average the satellite-transmitted phone calls cost as little as four cents per minute. Thirdly, the VSAT system is efficient. Satellite offers a coupling of real time and batch processing in a bandwidth efficient way. And finally, it is highly reliable.

In addition to these advantages mentioned by Wal-Mart's engineers, VSAT networks have some benefits over terrestrial networks. Satellite transmission is not distance-sensitive, and therefore is ideally suited to carry traffic over long distances to many different locations. Another advantage is that VSAT network management systems are easier to use than systems that manage terrestrial networks. The latter usually involves both the interexchange carrier and the local exchange carrier. With a VSAT system, users can employ a single network management system.

How important is the satellite network to Wal-Mart's operation? According to Allen, Wal-Mart stores are heavily dependent on it: "They would not be able to provide the dynamic services with the price pointing without it. The way that Wal-Mart does business is tremendously benefited by the addition of a real time satellite communication system."

In an extremely fast-changing environment, technologies are easily outdated. Wal-Mart is aware of the technological obsolescence. Although already harboring the most advanced technologies, the company is still looking for better technologies. Wal-Mart recognizes that its competitive edge provided by VSAT is only short-term. It would like to explore more opportunities on the technological frontier. The company has an interest in integrated services digital network (ISDN), believed by its engineers as the stepping stone to the future.

Wal-Mart's VSAT network also greatly facilitates the company's ability to operate the most sophisticated inventory control system in the industry. Wal-Mart practices the manufacturing industry's just-in-time concept for its inventory control. The VSAT network, the company's high-speed computer system which links virtually all the stores to headquarters and the company's 17 distribution centers, electronically logs every item sold at the checkout counter, automatically keeps the warehouses informed of merchandise to be ordered, and directs the flow of goods to the stores and even to the proper shelves.

Such in-depth inventory control helps detect sales trends quickly and speeds up market reaction time substantially. Furthermore, the automatic replenishment system reduces storage room and increases merchandise turnover and reduced costs. Wal-Mart has so fine-tuned communications between its store, suppliers, distribution centers, and its trucking fleet, that to resupply a store with a product takes only a day. Wal-Mart stores get 75 percent of their merchandise through the company's 17 distribution centers, thus eliminating middleman costs. The automated communications network helps Wal-Mart keep distribution costs at about half of those of most other chains.

"MAQUILADORAS": DOING BUSINESS ACROSS THE BORDER

North American Partnering

One use of business applications of high priority in this era of new trading blocks like Asia or the Common Market is the linking of countries on the American continent. Doing business with Mexico has vastly increased with the coming of "maquiladoras," or firms in Mexico that assemble or finish goods shipped in pieces or parts from the United States and then sent back to the United States for sales and shipping. The advantage lies in the low-cost Mexican labor and in Mexico's duty charges that are based only on the value added by the business operation. Most of these firms are jointly owned with a U.S. (and now some Japanese) company partners. Of the 1,450 maquiladora plants in Mexico, most are located in border cities. The maquila industry has replaced tourism as Mexico's second source of income. These production-sharing operations include a large variety of sectors, of which electronics, textiles, and coupon sorting are the most numerous. Partnering with Mexico may become a highly significant prospect if we enter into a free trade agreement along the lines of the one earlier executed with Canada.

"Twin Plant" Manufacturing with Mexico

In recent years, particularly as large companies with maquiladora operations have entered the picture, telecommunications has increased in importance for the efficiency of these operations. This is because the new generation of maquiladoras has many of the characteristics of the new industrial paradigm that includes such characteristics as just-in-time supply. This was introduced in Japan by automobile manufacturers and consists of reducing the parts in stock, thus externalizing the inventory costs. Another characteristic is "small-batch manufacturing," where the production is more specialized and requires flexible machinery (including robotics) and skilled labor.

The foregoing requires that the maquila site have excellent telecommunications links with its parent company, typically located in the United States. The transborder nature of this communication has been a constant problem for maquila operations, not only in antiquated transborder regulatory agreements, but in securing prompt and reliable service from the Mexican telephone company (Telefonos de Mexico or "Telmex," for short).

Several summaries from case studies by Barrera (1988) illustrate the uses of transborder telecommunications for business.

Fisher-Price: Using the Public Network

Fisher-Price, one of the world's largest toy manufacturers, is a subsidiary of Quaker Oats. It has an assembly plant in Matamoros, in the Mexican state of Tamaulipas. Its main offices are just across the border in Brownsville, Texas. This plant has been doing real-time computing with the mainframe located in Buffalo, New York, since December 1984. This is one of the companies that has been using the services provided by Telmex and has two sets of data lines, one of them as a backup.

The terminal in Matamoros is connected with the two data lines to the Brownsville offices and from there to Buffalo, New York. This firm has a comprehensive master data system called Master Production System which is comprised of three systems: the PLR, which deals with personnel matters; the MCS, which is the production-scheduling system; and the WYCS, which is an inventory system. The latter is a part of the corporation's Master Requirements Plan which gives the New York offices the capability for immediate purchasing of supplies. The Brownsville offices can link directly to offices located elsewhere, like the ones in Kentucky.

Telecommunications of this corporation is managed and operated by Electronic Data Systems, which is a subsidiary of GM. General Motors of Mexico still represents about 40 percent of the total work of the Gerencia de Servicios para la Industria Maquiladora of Telmex, an office which was created in 1985 to attend to the demand of the maquiladora industry in general. This office moved to Ciudad Juárez after a few months in Mexico City. The necessary cooperation between EDS and Telmex could become a moot point as the GM subsidiary has plans to bypass the Mexican network using microwave links and coaxial cable.

Zenith: Using Microwave

Zenith Electronics Corporation is one of the largest electronics firms in the United States. The principal research facilities are in Glenview, Illinois, with five manufacturing plants in the Chicago area and an assembly plant in Springfield, Missouri. Within the United States, Zenith has subsidiaries in Indiana (manufacturing television cabinets), Michigan (assembling microcomputer products), Texas (six locations serving as warehouses for materials moving between Mexico and the United States), and a variety of other locations throughout the United States established for display and sales.

Outside of the United States, Zenith owns a plant in Taiwan (manufacturing monochrome video displays and electronic components) and seven plants in Mexico. The plants in Mexico are located in Matamoros (manufacturing cathode-ray tube electron guns and other electronic components), Reynosa (manufacturing and assembling small-screen television receivers, color television chassis, and module boards), and Chinhuahua (one assembling cable products and one assembling power supplies). Two other plants are located in Ciudad Juárez.

Zenith is currently using two microwave systems for transborder communications, one in Reynosa–McAllen and another in El Paso–Ciudad Juárez. As of 1990, the firm is using a third one in Nogales–Tucson. The network configuration permits the exchange of data or other information between any two U.S. or Mexico locations, but it requires going through the corporate headquarters. The Taiwan facilities are isolated from the network and the exchange of information is done through the transportation of stored data. The firm is using the network to control the inventory, payroll, and production schedules of the plants in Mexico.

Westinghouse: Using Satellites

The Mexican operation of Westinghouse, named Sistemas Electrónicos Mexicanos, is not the typical maquila operation for two reasons: It is not an assembly plant, but a computer software producer; and it is not located on the border, but in Chihuahua. The location of maquila operations outside the border belt is an increasing trend, and cities such as Chihuahua, Monterrey, and Guadalajara are the sites of numerous plants. In the past, these plants faced more severe telecommunications problems than their border counterparts because they did not have the option of microwave systems and their telephone communications to border offices on the U.S. side involved expensive tolls.

Westinghouse was the first firm to solve this problem with satellite transmissions. They have been using the services from INTELSAT's IBS for a Chihuahua–Baltimore link. This link is a part of the Westinghouse Private Network and is used to send the software requirements and its architectural design from Baltimore, then developed in Chihuahua by graduates and interns from a nearby technical institute, and sent back to Baltimore for its installation and operation, both inside and outside the firm. This link is essential to the business operation and is used extensively. Besides data transmission, it is also used for voice communications, facsimile, and audio and video transmission.

THE "BYPASS" ISSUE

In many of the network examples in this chapter, large telecommunications users have bypassed the public network by installing their own microwave links, using satellite, or leasing lines from private providers. Flexibility and lower cost are often given as the reason for bypass. The problem this creates from a public utility standpoint is that when large users go off of the public network, so do their financial contributions to the public rate base. And since many large users require sophisticated applications, the pressure for service upgrades is also lessened on the public network. There is the potential for a negative to grow even more in this situation, as certificated telephone companies do not promote competitive advanced services to large users and these users, in turn, leave their network. The potential for public network upgrades, without pressure from large customers who have left, is then even less. Also, this leaves small businesses, who cannot afford to bypass, with fewer opportunities to upgrade their own telecommunications.

■   ■   ■

There seems little doubt but that aggressive use of information technologies is a key to U.S. economic competitiveness in the growing global economy. Investments in intelligent network applications may be our edge. But will these investments be mainly in the private sector and not available to all via a public network? This is a pressing policy issue.

5

■

Doing Business
on a Smaller Scale

Shirley Jay, who runs the local Century 21 office in Demopolis, Alabama, has many uses of telecommunications. First, she is on-line with a
national Century 21 computer, the CenturyNet. In addition to giving
access to national listings, the service allows real estate agents to have instant amortization tables and property financial statements. There are
numerous information services that allow Shirley to keep track of the
national and regional trends in the real estate business, including changes
in tax laws. Shirley has found her use of FAX steadily growing. She uses
it to receive descriptions of housing needs from people moving to Demopolis, who mainly consist of engineers for the paper mills and managers for other businesses in Demopolis. Also, she receives benefits from
Century 21's national television advertising. This brings customers
through name identification with her service and company. Jay pays for
these national services through a franchise fee and a monthly computer
charge.[1]

In this second business chapter, we concentrate on how the small business operator can stay competitive with large business or branch businesses through the use of information technologies. Shirley, in the
opening anecdote, had the benefit of belonging to a large organization.
But many small businesses do not have such advantages and cannot remain competitive unless they employ telecommunications-based services, which unfortunately some cannot have access to in small towns.

■   ■   ■

NEW OPTIONS FOR SMALL BUSINESSES

Telecom Strategies for Small Businesses

Nobody agrees on the exact demarcation between large and small businesses and being exact is not all that important to our discussion. We are mainly concerned with businesses that range all the way from a one- or two-person retail or service operation to a small manufacturer whose work force is under 100 people. These are typically locally owned and some are businesses found in small towns or rural areas in addition to cities. With some exceptions (like pulp mills), large businesses tend to be concentrated in or around cities.

Small businesses are benefiting from the advantages of expanded network services, some of which are summarized in Table 5.1. One of the best examples of the spatial expansion of markets in retailing is the use of outbound "WATS" calling for telemarketing and inbound "800 toll-free" numbers offered for customer inquiries and orders. These long-distance services are one of the largest growth areas in contemporary business expansion. Small manufacturers or retailers serving widely dispersed

Table 5.1. Applications for Small Businesses

Current and Growing

—Telemarketing using wide area telephone service
—Toll-free (800) calling for customer inquiries, orders, or troubleshooting
—Customers order by FAX
—FAX links with suppliers
—"Batching" orders via computer and modem
—Some on-line accounting or personnel management services
—Remote management, consulting, and inventory assistance

On the Planning Boards

—Special ISDN packages for small businesses, including customer premise integrated voice, data and FAX equipment
—Public FAX and data communication "booths"
—Access via the network to computer-assisted design, computer-assisted manufacturing, and computer-assisted engineering services
—Access to satellite-based transportation tracking services
—Access via the intelligent network to a full range of managerial and consulting services
—Multimedia electronic catalog selling and shopping

"niche" markets can gain an economy of scale by using telemarketing methods. Another example is the small manufacturer, perhaps a "supplier" with only one or several large customers. In this case, direct data links with the customer can be used for ordering, inventorying, and shipping control. In fact, many large corporations will not do business with a supplier unless they can go on-line.

There are also small retailers whose clientele may be concentrated locally but whose main problem is dealing with a wide variety of suppliers and a vast inventory of small items. In such cases, as with drug or hardware stores, it is becoming popular to engage network services for managing or ordering inventory. In some current services, the merchant keeps inventory lists in an in-store computer that daily communicates with a wholesaler who automatically restocks items, adjust prices, adds special orders, and can give the retailer an analysis of sales by shelf space. Pharmacists can use such services to keep customer profiles, generate reminders of prescriptions that need refilling, check for possible drug interactions, and automatically process insurance billing.

Rather than looking at small firms in terms of types of businesses, we can examine them in categories that highlight the telecommunications uses in their business operations. This is a focus on how they do the business rather than their end product. The key factor for developing telecommunications applications is an understanding of the "communications interfaces" in the operation of these businesses. The following examples drawn from a field study (Williams, Sawhney, and Brackenridge, 1990) provide a broad framework for developing an understanding of these interfaces. The categories include: (1) widely dispersed customers, (2) a few large buyers or suppliers, and (3) localized businesses.

(1) Widely Dispersed Customers

The businesses that fall into this category have one thing in common, namely the need to communicate with customers distributed over a large geographic area. These businessses tend to serve specialized markets which are too dispersed to support a concentrated distribution channel. Typically they tend to be specialized light engineering operations which have their customer base spread over the continental United States. Many use WATS services to maintain contact with their customers and the United Parcel Service (UPS), or another shipper, to deliver the goods. It is the WATS-UPS combination which makes it possible for many of these businesses to operate out of small communities in serving widely dispersed "niche" markets.

Two rural examples we have studied include Prodeva, Inc., of Jackson Center, Ohio, the opening example in Chapter 2, which manufactures blades for plastic shredders (recycling industry) or the Berlin Co., near Lake Milton, Ohio, which produces medical products mainly for wholesale distribution to trainers or veterinarians who specialize in horses. Telecommunications, mainly in the form of toll-free calling services, appears very valuable in meeting these needs. Using WATS for outbound marketing and customer troubleshooting, or giving customers an inbound 800 number, has changed the nature of many of these niche businesses. As one rurally located manufacturer told us, "give me my 800 numbers and United Parcel Service, and I can do business anywhere."[2]

(2) A Few Large Buyers or Suppliers

Although the businesses in this category are themselves small, they usually have to deal with a few large buyers or suppliers. It is the need to interact with large entities which shapes the communications reality of these businesses. Unlike the first category, the communications problem is not that of accessing isolated pockets of demand but interfacing with the sophisticated systems of their buyers or suppliers. The development of their communications linkages is often mandated by their larger business partners. An excellent example of this are the pharmaceutical suppliers that refuse to deal with pharmacies which do not have on-line ordering capabilities. In most cases, "on-line" means being able to connect a small computer via a modem to the telephone network.

The largest application of on-line communications in small retailers we have studied tends to be pharmacies or hardware stores. Also for small manufacturers, there have been cases like that of a General Motors supplier who bought a modem because it was mandated by the company. Most of these examples use a batch transfer once or several times a day to do ordering, billing, or reporting inventory changes. This exchange of information may sometimes by augmented by use of FAX or voice messages. It is not difficult to identify the next generation of on-line users, most of whom are now using FAX. They may find themselves printing information out of a small computer in order to FAX it, while the recipient types the information into a computer. It is a natural next step to link the computers directly.

(3) Concentration in the Local Market

This category of businesses concentrates its activities in a town, part of a city, or in a regional area. Shirley Jay's (opening example) real estate business represented one type of firm that uses national information over a network, but also uses voice and FAX intensively in the community. There are also many cases of uses of more local networks, as was the case with Oberfield's Inc., a concrete block manufacturer, who produces and serves the region within a radius of 100 miles around Sunbury, Ohio. Transportation and long-time local business relationships generally define the geographic range of this type of business. Voice telephone, including mobile radio, is used extensively, but the latter may eventually be replaced by cellular telephone. The biggest problem with voice for this company is the toll rates involved within the territory it serves. Although its three plants are in geographic proximity, they fall within the territories of three different Ohio local exchange companies, namely Sunbury (United), Delaware (GTE), and Columbus (Ohio Bell). Cellular telephones will probably offer cheaper local calling if the entire area use is taken into account.

SNAPSHOTS OF SMALL BUSINESSES

Two Drugstores: On-Line Management Makes the Difference

Today's customer wants services from a pharmacy at all hours. They like efficiency; they don't want to drive or walk far, and they appreciate getting credit when they need it. Older folks like to be able to call in their prescriptions. Often customers might want to pick up a few items besides a prescription, like toothpaste, film, candy, or a last-minute birthday gift. But not all pharmacies are surviving, as is illustrated in the contrast between two stores, not more than a mile apart in two Bergen county boroughs in northern New Jersey (ongoing unpublished project).

One store, which to avoid embarrassment we'll call "City Pharmacy," is located near the corner of what once was this small borough's main intersection, near a bank, diner, barber, and a bagel shop. Inside, the store is friendly but you get a 1950s feeling (and if you weren't alive in 1950, think of the old TV show "Happy Days") as if there still might be a soda fountain. (You note that some of the tubes of Vicks Vaporub have a package style that was replaced about two years ago.) The owner, Mannie, has indeed operated the pharmacy since the fifties and his customers have been several generations of families that have probably seldom bought a prescription anywhere else. Mannie explains that in the good old days, most neighborhood physicians automatically wrote or called prescriptions to him unless the customer requested otherwise.

But times are not so good anymore. The longtime families are being replaced by new move-ins who do most of their shopping at malls. If some of the newcomers do come in, they don't want to pay cash but want Mannie to charge their prescription to their insurance plan (which he tries to avoid because then he won't get his money for another month), or their kids come in to buy candy (a longtime tradition, but now a money-losing one). These people also complain about the lack of parking spaces, which amazes Mannie because they drive from probably no more than three to five blocks away. Mannie will deliver prescriptions but he usually can't get out before 3:30 in the afternoon when his part-time clerk gets out of school for the day. If you can engage Mannie in an informal conversation, he'll tell you that when he retires, the store will probably go out of business because he just can't show a healthy cash flow, and the main thing of value is the building which he bought over a period of 30 years. He has been contacted by a video rental company that might be interested in the location.

"Rite Value Pharmacy" is in a similar borough, but about two miles north and where a strip mall rather than a small "downtown" has developed. Sally, the store manager and part-owner, will explain that the three most important keys to success in this business are "convenience, convenience, and convenience." People want to get "walk-in" prescriptions immediately; they want you to bill their insurance directly, and if they have a sick kid with a fever, they want the medicine delivered within the hour. They want easy and free parking and will favor stops where they can do most of their daily shopping at once—groceries, cleaning, maybe even fast food. Sally explains that most of her customers, men and women, work all day, so much of the business activity is after five in the afternoon. "Sure, they like a friendly smile, but this business isn't so personal anymore," explains Sally. "They're not here to talk; they want to get their prescription, film, or personal items, and get on their way."

If you can get Sally talking about her store—and that's hard to do because she's usually pretty busy—she'll tell you that her success is not only due to the usual professional background a druggist must have, but you also have to know modern retail store management methods. "There's part of it," Sally emphasizes, pointing to a personal computer just within reach of where she fills prescriptions. She describes how her store is part of a cooperative buying and management service most of which she operates with her computer, both in a stand-alone mode, as well as being linked via the phone line to a mainframe computer in Chicago.

Each night the computer automatically dials and logs onto the service's host computer to accomplish the following: (1) updating prices on all pharmaceuticals, (2) placing restocking orders that Sally has entered into the computer during the day, (3) checking inventory records and making suggestions as to what items Sally might be low on and has not reordered, (4) making seasonal suggestions for ordering when relevant ("reminds me that hay fever season is coming on"), (5) updating a summary of her fastest- and slowest-moving items ("any nonpharmaceutical that sits around more than 90 days is dropped from my inventory"), (6) receiving third-party insurance billing for the day, (7) calculating and presenting a cash flow analysis for the day ("my commercial bank account is usually credited the next day"), and (8) issuing any bulletins on information that might be important to the business ("an insurance company might have bad credit"). During the day, the same computer holds Sally's customer records, auto- matically logs all point-of-sale information via her cash register terminals, and will warn if a customer might be ordering a drug that will interact with one already being taken.

"Has being a pharmacist lost its edge to impersonal machines?" one might ask of Sally. She is quick to explain that it is the routine clerical items that are automated, which leaves her more time to think and plan, and to give customers personal attention when they need it. But there still may be little choice in the matter. If she can't provide the convenience to her customers, the pharmacy at Wal-Mart, Shop Rite Supermarket or discoun- ters who sell via 800 numbers will take over the business, Sally explains. Ironically, Sally did not learn about computer-assisted management in her pharmacy degree program; she learned about it during her first year out when working for a large chain operation. The rest of her management education came "hands on" with the equipment in her new store.

Telecommunications delivers to Sally's small business most of the mod- ern management of her largest competitors. You see these services in- creasingly in small retailers who must handle large inventories like hardware and drug stores. And, of course, you see "companywide" computer- assisted management systems in such chain-store operations, and even now in fast food chains like Wendy's or McDonald's.

Building Supplies: A Case of Mini-Networking

Some small business people develop their own network services as in the case of Kim Mayton, an innovative Demopolis, Alabama, businessman, who planned at the outset to use a computer network to manage his three- store building supply business in rural Alabama. Headquarters is the Demopolis store located on highway 80, next to Wal-Mart, on the south edge of town. Mayton's building supply company is an outgrowth of the family's concrete business which was established by his father. Mayton opened his own building supply store in 1975. The business was so successful that by the mid-1980s it had grown beyond its original capacity, and he saw the need to have multiple sales outlets. "But why duplicate the managerial tools at each store?" questioned Mayton. "We needed to computerize and link the system among the multiple outlets. Even with one store, we felt we did not have enough control of our business. Our people couldn't get work done fast enough to meet demands and we constantly made clerical mistakes in inventory, accounting, and credits. At that time, we realized that we had outgrown our present system." And so Mayton expanded his business outlets and linked them on a computer network over leased telephone lines.

In the headquarters store and each of his branch outlets, clerks use computer terminals as they serve customers. With only a few key strokes, they can tell a customer all the information about a product. If the customer decides to buy that product, the clerk will press a few more keys and then the transaction is done. The customer either pays in cash or receives a bill calculated by the computer at the end of the month. The computer not only allows on-line transactions but offers a range of other services. Through his computer and store network, Mayton monitors his inventory, issues monthly statements, balances his accounts, and, now and then, sends electronic messages to his employees in other branches.

Building supply is an inventory-intensive business. Mayton presumed that computers, with their ability to process data in no time and a suitable memory capacity, would be a solution to his problem. With little knowledge about computers and telecommunications, Mayton began his investigation by reading trade journals and talking with people who already used computers. After extensive traveling out of town and visiting other building supply businesses, he found out that in larger cities almost everybody with the same volume of business as he did had installed computers.

Having gathered information from his out-of-town predecessors and from computer magazines, in 1985 Mayton wired his office. He selected a package offered by Triad, a Livermore, California, based firm specializing in building supply business computer packages. Mayton rewired his office in the current location when he moved the store from downtown two years ago.

While telecommunications enables his next-door neighbor Wal-Mart, the megamillion discount giant, to run a national business, it allows Mayton, a small-town businessman, to manage a three-store chain in a much smaller radius. Shortly after his office was wired, Mayton opened his first branch store in Greensboro, a small community 25 miles north of Demopolis, in 1986. Next year, he established another shop in

Thomasville, 50 miles south of the home office site. To link his computer with these two stores, Mayton leases a dedicated phone line for on-line data transmission between the Demopolis home office and each of the two branches. His computer network will make it possible for Mayton to expand his business to still more locations. Mayton said he would not have opened the two branches if he had not had a computer. "With the home office computer supporting the inventory and accounting functions, all it takes to open a branch store is to hire a few clerks to wait on customers." At the time of this study, Mayton planned to open a fourth store the following spring.

In addition to his computer network, Mayton relies on voice telephone and FAX to coordinate his stores. He calls each of his branches several times a day and uses FAX to send documents and letters to and from them. FAX is also used to place orders to vendors and receive orders from some of his customers. Mayton said FAX is used on a daily basis and the traffic is getting busier every day. Although the computer has the capability to send electronic messages, Mayton does not use this function often because "it is much easier just to pick up the phone." Mayton's store is the only building supply firm in Demopolis with a computerized operation and one of the few small businesses with computers. This has given him an edge over his local competition.

Mayton said the computer network benefits his business in three ways. First, it helps the store operate more efficiently with less inventory. "The computer keeps a 'selling history' from which we know what sells and what doesn't. Therefore we don't need to keep a large inventory to meet customers' needs." Second, the computer helps the store have better control over accounts receivable. "It keeps track of our customers' paying records so we know who is paying and who is not—and we will contact those who don't pay." Mayton said after the computer network was installed, the average age of his accounts receivable has dropped signifi-cantly. Third, the computer network has increased the business's profit. "It raises our profit margin; we are growing at about 2 percent a year."

Being next to Wal-Mart, Mayton's store benefits from the busy traffic that the discount giant draws. Although Mayton considers Wal-Mart a competitor, his store and Wal-Mart seem to be able to coexist well. Despite the general animosity toward Wal-Mart among small businesses in Demop-olis, Mayton credits Wal-Mart for stimulating competition. "Wal-Mart forces you to be a better retailer. It forces you to use ads more effectively and become more competitive." Mayton said service is their strong suit over Wal-Mart. "We deliver and service. We also give customers consultations." Mayton said many small stores cannot compete with Wal-Mart "because they don't have technology."

"We do, and it has paid off."

Insurance: "The Phone's My Secret Weapon"

Highly personalized small businesses can use telecommunications to expand their market coverage as was the case with a small insurance firm included in one of our studies (Williams, Sawhney, and Brackenridge, 1990). Because the owner of this small-town insurance agency wished to remain anonymous, we'll use the name "Sheila" and call the town "River City."

"This is a very person-to-person business," explains Sheila. "You've got to get your client to remember your name and phone whenever they think about insurance. That's why I moved heaven and earth to keep my old number when we changed out the phone system last year, and that's why I use call forwarding from that number to reach my cellular phone when I'm on the road," she adds.

Sheila has been an independent agent for about 15 years, picking up from a business her dad started on a part-time basis in this rural Western town of 2,400 about ten miles off the interstate highway. "But I had to turn it into a full-time living because I have two children to support," Sheila explains, as she fills in the details of her success story. She describes how many rural people in this business are part-timers who maybe only write one type of policy, say, "life." They get the business from their small-town friends and neighbors whom they mostly see face-to-face; they communicate by mail with their underwriters.

"I'm maybe a little sorry to say that I've built my business by taking over from many of these part-timers, but I can truly offer my clients a lot more for their money," she says. "I can check out the details of a new policy or file a claim almost instantly. I do this with 800 toll-free numbers to my underwriters, most of whom use FAX when helpful for details, and with several of whom I now trade information via my desktop computer hooked to the telephone," she explains. "In fact, for two of the companies I can print out a finished contract on my laser printer with my computer hooked into the phone line." This, plus having an office on the town square, has brought her the bulk of the in-town business. "But to get the size business so I can earn a living, I have also expanded into five nearby towns," Sheila continues. "If folks can't call me toll-free from their town, I buy a 'foreign exchange' number that works like a local one." Sheila explains that people don't like to make toll calls, even if she will pay "collect" for them. "They just don't like the bother, and a few have said it makes them feel cheap to ask the operator to have somebody else pay the 60 cents. I'm making a good living now as an independent agent because I've brought modern business methods to my office."

"The phone's my secret weapon."

Dental Chemicals: Going to Direct Marketing

Rick Meese mixes chemicals in a small plant located in Carpenteria, California. His main product is a film-developing formula for dental X-rays that his father patented over 30 years ago. Until about five years ago, Rick sold his developer through professional salesmen who called on dentists to market a broad line of products—everything from dental tools to filling materials to X-ray equipment and supplies. This arrangement took 20 to 40 percent off the top of the final price of his product. When he computerized his accounts, Rick saw that about 80 percent of his business came from about 20 percent of his customers, mainly dentists operating large group practices. Also he knew many of these customers personally because he met them or their representatives at annual conferences on dental products.

At the risk of losing customers who had a close relationship with the field salespeople, Rick decided to go to direct marketing. He set up WATS calling to lessen the costs of calling his customers, and gave them an 800 toll-free number to place their orders directly (and included a discount for doing so). He also guaranteed one-day shipping. Although he lost a few of his customers, his best stayed with him, which led Rick to sponsor a dinner each year at the convention. This direct contact not only cut out the middleman and the costs, but from his more personal contact with his best customers, Rick obtained ideas for several new product lines.

By using telemarketing methods, Rick retained most of his best customers, got new product ideas, and cut his costs of doing business by an average of about 20 percent (including costs of toll calling). In the last several years, voice call orders from his biggest customers are being replaced by FAX.

Rural Manufacturing: Telemarketing Gives an Edge

Lacal, in Jackson Center, Ohio, provides an interesting example of how a small manufacturer can have the advantages of a rural-based business (low space and labor costs) yet can use telecommunications to reach a wide market. Lacal manufactures cutting blades for agricultural implements. It keeps in touch with its widely dispersed clientele through periodic tele-phone calls. On a six-week cycle, Lacal calls its customers to see if they have a need for a cutting blade which is then shipped via UPS. Lacal has nine full-time telemarketing employees who, on a daily basis, make as many as 300 outbound long-distance calls and receive about 200 inbound long-distance calls. Its monthly telephone bill is in the range of $12,000 to $15,000. Among its challenges, Lacal had to develop its telemarketing through trial and error. In its rural location, there were no readily available consultants to help plan telecommunications applications.

Multinational Manufacturing: "Reverse" Maquila

Small manufacturers are also decentralizing their operations, including international operations. Maverick Arms, a subsidiary of O. F. Mossberg, a major gun-making firm headquartered in Connecticut, makes its gun barrels in Mexico, the firing mechanism in Connecticut, and assembles the final product in Eagle Pass, Texas. Attracting Mossberg to Eagle Pass was a success story for Central Power and Light (CP&L), which fields an economic-development staff in its service areas.

None of this would be possible explains Gabriel Bustamante, the assist-ant plant manager, if it were not for modern communications. "We have to coordinate manufacturing and shipping among three diverse locations, one being across the border." Because the Mexican government will not allow the assembly of guns in northern Mexico, Mossberg was seeking a "reverse maquila" type of arrangement. The company established ties with a group in Torreón, Mexico, which acted as a representative to the Mexican government. Raw materials arrive from Connecticut; shotgun barrels are manufactured in Torreón, Mexico; trigger assemblies are manufactured in Connecticut; assembly, finishing, and shipping are centered in Eagle Pass.

Manager Gabriel Bustamante, a resident of Piedras Negras just across the river in Mexico, trained as a civil engineer at Texas A&M University, commutes across the border every day. He has worked in maquiladoras for the past six years in Ciudad Acuña and Monterrey before returning to this home in Piedras Negras to participate in the border development there. Bustamante reports that Telefonos de Mexico (Telmex) service is adequate for communications in Mexico, but could be better in terms of service reliability and the ability to make upgrades in a short amount of time. Networking would be easier if he could deal with one major telecom-munications provider instead of three—Telefonos in Mexico, AT&T for transborder links, and Southwestern Bell for local service in Eagle Pass.

Feedlots: Bringing High Tech to Agriculture

Global competition has been introduced to Maverick County, Texas, once known for its year-round agricultural production. In 1989, the county produced $84 million in commodities, 90 percent related to the cattle business. Large feedlot operators such as Alta Verde Industries, Inc., which employs about 300 people, bring in bonded cattle from Mexico, fatten them, and then send them back to Mexico where American-fed beef brings high prices. However, the industry is mature and tapering off, largely due to the elimination of tax loopholes in 1986.

Telecommunications link the feedlot operations to their markets and customers. Staying in business means keeping current with prices and maintaining close and instant communication with clients. Alta Verde must use a satellite service, on-line computer service, and FAX to stay competitive with other feedlot enterprises. FAX is used for customer correspondence, banking relations, advertising, and contracting; the company has about 50 customers on its FAX mailing list. Additionally, the company uses in-house desktop publishing to compose its newsletters. Of all the technologies, the satellite receiving system is probably the most valuable because of the specialized news service it provides through the Texas Cattle Feeders Association (TCFA). This news service, created by the Bonneville Network, carries U.S. Department of Agriculture (USDA) information as well as futures and livestock price quotations from the Chicago Board of Trade and Chicago Merchants Association, among others. The TCFA also uses the service to provide an electronic bulletin board of such things as notices of cattle sales and association meetings. Alta Verde sends daily sales information and receives hourly postings of market prices, which saves it endless calling.[3]

Telemarketing: "800 Capitol of the World"

One of the more innovative uses of telecommunications for small businesses is the expansion of their markets through telemarketing. Such usage is popular in Nebraska. In fact, Omaha has been called the "800 Capital of the World." Omaha and the surrounding area are populated by telemarketing businesses that are intensive users of 800 and WATS services. These businesses employ a large number of Omaha's work force. The Omaha Chamber of Commerce lists over 60 firms in that city that are involved in telemarketing. Of these businesses, some are direct telemarketers whose daily business involves long-distance telephone sales. Some are contractors who offer their telemarketing expertise to businesses that are not interested in setting up such activities in-house. Several are reservation centers for large business chains such as hotels. A few provide support services such as software and record keeping or number crunching via telecommunications links.

Many firms have chosen to locate in the Omaha area for several reasons: the characteristics of the work force—Midwestern "flat" accent and work ethic[4]—the state's central location in the United States, and the support services offered from Nebraska's telecommunications industries (Schmandt et al., 1991).

Although telemarketing may have brought new job opportunities to mid-America, it is important to note that most employees are part-time housewives or students. The implication is that developing telemarketing services as a local business may not have the financial impact as a light manufacturing industry, modern agriculture, a shipping business, or a large retailer. We refer to a telemarketing operation itself, not the use of telemarketing by a local business to expand its markets.

Telemarketers can pull up stakes quickly, given the attraction of a lower cost and willing labor force located elsewhere. This is not to discourage telemarketing as an area of business development, but as a business unto itself, it may not have all of the glitter that some have promised for the economic revival of mid-America (Strover and Williams, 1990).

TELECOMMUNICATIONS AND
THE DILEMMA OF SMALL BUSINESSES

If there is any one group that has suffered the most loss of opportunity at the hands of divestiture, it is American small businesses. Although it is easy to locate small business success stories in uses of telecommunications to improve management methods, to enlarge markets, to link efficiently to suppliers, to serve as a key supplier, or to coordinate branch outlets, getting those uses up and going is a real challenge in most cases. The problem is that in the attempt to bring competition to the telecommunications marketplace, most small business persons are forced to deal with multiple equipment and service providers. It is left to the business owners, lest they hire an expensive consultant, to "put the package together." The local exchange company sells access to the local network; another company will likely offer better prices on premises equipment; still another may be needed for computer software and linkage of machines on a local network; then added to these are the competing long-distance providers. The small business person will usually not have telecommunications expertise, and has little, if any, time to do telecommunications planning and to evaluate decisions. As we have been told by numerous telecommunications marketing representatives, it is difficult to attract small business persons to regional seminars because they simply cannot afford to be away from their enterprise for a day or two. They may be the only person who can run the business.

In the days of the Bell system, a small business person could have the benefits of "one-stop" shopping; that is, one representative could serve for all of the above needs. But today, between the restrictions put upon local Bell companies to engage in multiple lines of business and the cost-cutting they have had to implement, this country's largest local exchange service providers cannot serve as a sole contact for telecommunications needs, let alone offer much free consulting time. In rural areas, sales representatives will seldom travel to a customer site unless it is a major order. They cannot afford the time that sometimes involves two to three hours of driving relative to an hour with a customer; sales quotas discourage spending time and energy on small, and especially remote, customers. This leaves the "telecommunications edge" to the large businesses, including the many chains that compete with locally owned businesses. Perhaps the only way around this dilemma is to create, through regulatory initiatives, more incentives for local exchange companies to assist small business development. Additional help could come from more inclusion of telecommunications applications strategies in business or trade school curricula, training a new breed of small business telecommunications consultants, and including more practical programs on telecommunications applications in trade associations involving small business membership.

■　■　■

U.S. small businesses are the nation's largest employer and the opportunity to open one's own business is a longtime American dream. It is unfortunate that this sector of the economy may have suffered the most at the hands of divestiture of the Bell system. Hopefully, regulatory and developmental efforts in the coming years will bring intelligent network services to this sector.

6

·

Promoting Science
and Technology Transfer

You'll never meet a more enthusiastic person than Umberto Bozzo, especially in his role as director general of Tecnopolis Novus Ortus, a research park near Valenzano in Southern Italy. Because "tecnopolis" (or the English version "technopolis") is an intriguing label, Umberto usually launches into an explanation out of anticipation of his listener's predictable first question. He describes a technopolis as a "technology center," a "high-tech corridor" or "triangle," or, in short, an environment linking technology with economic development. The goals of Tecnopolis Novus Ortus, in Bozzo's terms, are not only to speed the development of advanced technology industries in Southern Italy (most have been in the Northern cities of Milan and Turin) but to interface this development directly with plans for the 1992 integration of the European Common Market (some 320 million citizens and a $4 trillion economy). Bozzo believes that the key infrastructure component for Tecnopolis Novus Ortus is the configuration of information technologies as available on the telecommunications network. This network not only serves as the central nervous system of the technopolis but is the basis for links and partnerships with R&D centers throughout the Common Market, and eventually the world. The network overcomes traditional barriers of time and distance; it provides near-infinite options for scientific cooperation as well as the transformation of inventions to products, or "technology transfer." No longer is the Apulian region confined to a vineyard industry. There are new options, new community structures. A research team can easily operate in what was an agricultural economy. At any time, scientists can closet themselves in a private workshop in the technopolis or can get on the network with colleagues in Paris. Or they might choose to retire to the Novus Ortus espresso bar for the evening.[1]

In this third chapter on business applications, we examine uses of networks to promote scientific activities and the bringing of new discoveries to market. Such undertakings may be a key basis for the economic development of post-industrial societies.

■　■　■

SCIENCE AS INDUSTRIAL POLICY

As in the Tecnopolis Novus Ortus example, the promotion of science as a basis for economic development policy is a global phenomenon among the world's industrial leaders. A common theme is the revitalization of traditional industries, development of research consortia, planning of industrial parks and even "science cities." Science is seen an increasingly important and renewable source of economic growth as in the development of new materials, products, processes, or treatments (see Kozmetsky, 1990). This could range as wide as the development of ceramic conductors, flat screen television, parallel computing, or gene splicing, to name a few examples. The point is that scientific invention—especially invention that is marketable—can be the cornerstone of a postindustrial economy. Current and new applications of telecommunications and computing not only facilitate the basic processes of scientific investigation and collaboration, but are also increasingly recognized as critical for "technology transfer." Major types of network services for science and technology transfer are summarized in Table 6.1; these are used to:

- support collaboration among scientists, scientific laboratories, and government agencies
- promote collaboration between industrial and university-based scientific laboratories.
- support the provision of continuing education in engineering and the sciences
- provide opportunities for research leading to the marketing of network products (including software) and services
- attract R&D or high tech industries to a city or region
- hasten the technology transfer process
- improve library service delivery via such networks
- enhance educational programs in universities and secondary schools, and eventually primary schools.

Table *6.1.* Network Services for Science and Technology Transfer

---

Typical Basic Services

—Electronic mail (messages from one person to another)
—Bulletins (messages from one person to many)
—News (messages from many to many)
—Conferences (conversations between a few or many)
—File transfer (movement of computer files from a source)
—Computation access
—Database access and search function
—Encryption (encoding) and security protection (passwords)
—Directory service (similar to while pages telephone book)
—Gateways or interconnections to other networks or databases

Advanced Services

—Video-conferencing (compressed video or full-scan video, one-way, two-way, multiways with video picture or audio-only response)
—Access to expert systems
—Computer graphics (computer-aided design, modeling, and other advanced graphics in color)
—Parallel computation among computers
—High-level computation requiring supercomputers
—Stimulation (e.g. weather, astronomy) requiring supercomputers

---

EXAMPLES OF MAJOR NETWORKS

Practical Goals

Especially important for scientific advancement are the computer networks[2] typically found in use in major research universities. These range from a wide variety of local area networks to memberships (or linkages into) inter-university networks like BITNET, and larger-scale, high-capacity networks like NSFNET. These networks permit researchers at colleges and universities, and industrial and government laboratories, to work simultaneously on a common body of knowledge. This increases the productivity of scholars, hastens the pace of scientific discovery, and accelerates the transfer of research results. Roberts (1988) describes national and international networking goals as to:

- increase research productivity by improving access to information, to supercomputers, and to other specialized sources
- advance the quality of academic research and instruction by expanding opportunities for collaboration and cooperation

- shorten the time required to transmit basic research results from campuses to the private sector and thus enhance national research and product-development capability
- broaden the distribution of scholarly opportunity and creativity by connecting faculty, students, and staff from diverse geographies.

We next examine several examples of scientific networks.

Internet

Internet, a worldwide "master" internetwork in 1990 of some 700 local, regional, national, and international networks, links 500,000 users at university, industry, and government research sites. The network is expanding at a rapid rate: 20,000 computers connected in 1987 grew to 60,000 in 1988. Internet acts both as a facility to share resources between organizations and as a test bed for new innovations in networking. The oldest network in Internet is ARPANET, the U.S. Department of Defense Advance Research Project Agency's (DARPA) experimental packet-switched network, introduced in 1969. ARPANET was due for phasing out in 1990. Its traffic is to be carried by MILNET and NSFNET (National Science Foundation Network). NSFNET, one of Internet's major constituents, is now and will become even more important to advanced research activities. Consequently, this network was chosen from dozens of research-oriented networks to illustrate technology transfer in action in this chapter.

Close-Up of NSFNET

NSFNET is a national network created to improve communications, collaboration, and resource-sharing in the science and research community. Figure 6.1 shows NSFNET's major nodes. NSFNET interconnects major regional networks at 250 universities and research sites and provides any recipient of an NSF grant with access to a supercomputer. The network is structured on three levels:

- backbone consisting of six NSF-sponsored supercomputer centers interconnected on long-haul, very high-capacity trunk lines
- mid-level networks connected to the backbone and to other mid-level networks and international networks
- campus networks

The NSFNET backbone uses MCI's fiber optic circuit and digital microwave radio network to carry data. The management and operation of

Figure *6.1.* NSFNET's Major Nodes

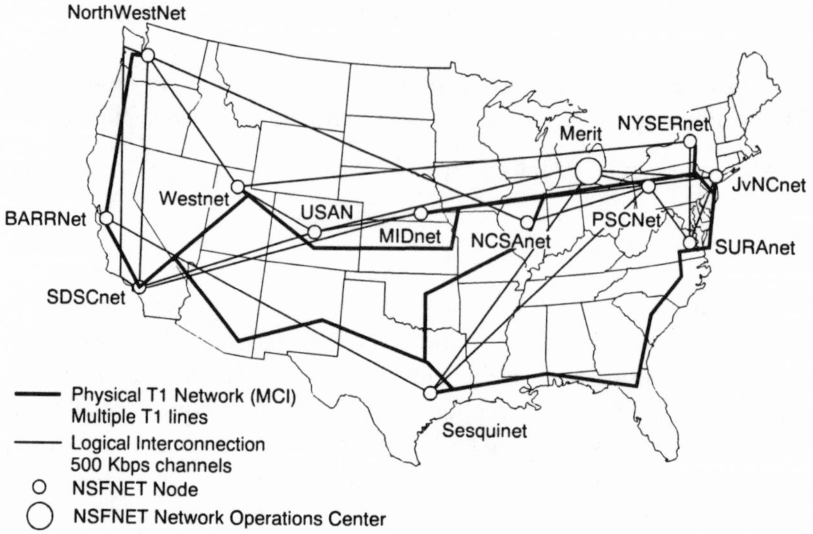

NSFNET's backbone is supervised by MERIT, Inc., a nonprofit consortium composed of eight Michigan universities (LaQuey, 1989).

NSF's Mid-Level

Eight mid-level networks are independent entities in a federation linked to the NSFNET backbone. An additional eight networks complete the mid-level group. Among the NSFNET international networks connected through mid-level members are EASINET (European Academic Supercomputer Initiative), JANET (Joint Academic Network, United Kingdom), and JUNET in Japan.

BARRNet operates on the second level of the NSFNET and connects universities and research organizations in northern California. The network was expected to have 40 members at the end of 1989 and over 60 at the end of 1990. Universities (such as Stanford and four University of California northern campuses) and industrial research members (such as SRI International, Hewlett-Packard, Xerox PARC, and Apple Computer) share the network with several government and private research laboratories (such as Lawrence Livermore National Laboratory, NASA Ames, and Monterrey Bay Aquarium Research Institute); (Baer, 1989).

THEnet (Texas Higher Education Network) provides service statewide and in Mexico to 50 members, including universities (such as the 17 campuses of the University of Texas system, Texas A&M, Rice University), industrial research members (such as Lockheed, Schlumberger, Texas Instruments) and other organizations such as Sematech, Microelectronics and Computer Technology Corporation, and Superconducting Super Collider Laboratory. THEnet functions in a joint cooperative effort with Sesquinet, a mid-level network connected to the NSFNET backbone. The Texas networks combined form one of the largest of the regional networks and include 2,000 nodes.

These and other networks will play increasingly important roles in fostering interinstitutional exchange and the diffusion of ideas, of products, and of innovative procedures. Network building is a dynamic, growing endeavor. Over the years connectivity will improve, while speed and capabilities will increase.

NYSERnet

One of the six major centers associated with NSFNET is the New York State Education and Research Network (NYSERnet), a high-speed data communications network linking universities, industrial research laboratories, and government facilities in New York State. The goal is to give greater access to "computing and information resources which will aid in improving economic competitiveness" (NYSERnet, Inc., 1987). NYSERnet is a nonprofit company formed in 1985 by a group of New York educators, researchers, industrialists, and NSF. Among the 47 users of NYSERnet are Columbia University, New York University, Polytechnic University, and Cornell University, as well as IBM, Kodak, and Brookhaven National Laboratory. The Cornell National Supercomputer Center and the NorthEast Parallel Architecture Center supply electronic library access and additional network services to NYSERnet.

NYSERnet was the first network of its type to involve facilities owned by telephone companies, in this case, NYTel and Rochester Telephone. By 1989, however, all the facilities were owned, operated, and maintained by NYSERnet, who leases lines from the common carriers as would any other customer. NYSERnet's revenue stream is 10 percent supported by grants from New York State and the National Science Foundation. The remaining 90 percent is funded by users of NYSERnet who pay annual fees from $15,000 to $78,000 (depending upon speed) to access the system (Williams and Brackenridge, 1990).

Projects promoting technology transfer are daily occurrences at NYSERnet. Four examples of technology transfer are the Apple Computer–University of Rochester connection; the Hartford Graduate Center–NorthEast Parallel Architecture Center connection; the Alfred

University–Cornell supercomputer connection; and the State University of New York at Buffalo–Ames Research Center connection.

Engineers at Apple Computer in Cupertino, California, conduct a sizable amount of development work in conjunction with the University of Rochester. Using BARRNet to connect to the Internet to connect to NYSERnet, Apple's researchers are able to do industrially relevant work sitting at their own computers in California.

The Hartford Graduate Center is currently deeply involved in researching artificial intelligence. Researchers desiring to test human depth perception for application to robotics devised a model of how a part of the human brain sorted out information. To test the theory, the Center needed a method to mimic the brain which could be provided by parallel processing. Using NYSERnet, the Hartford Graduate Center is able to log in to the NorthEast Parallel Architecture Center to conduct the appropriate tests.

Alfred University is a national center for ceramic engineering where researchers study ceramics for a variety of applications. In order to work with highly complex equations needed for developing models, the university connects with the Cornell National Supercomputer Center through the facilities of NYSERnet.

At the Computational Fluid Dynamics Laboratory at the State University of New York at Buffalo, research in the areas of combustion, computational fluid dynamics, and turbulence is being conducted. Through NYSERnet connections the lab has gained access to supercomputers at Cornell, the University of Illinois, and the NAS-supported Cray-2 at NASA Ames Research Center.

Another kind of technology transfer also can be found on the network. Users around the country have the advantage of public domain software. Users are able to "borrow" very sophisticated (and, otherwise, very expensive) programs whenever they need them by tapping into the network. This method of sharing enables even small-budget projects to perform at the same level as those who are in big-project operations. Small projects are often highly innovative and centered on a specific technological idea or breakthrough. It is critical to keep them supported.

SPECIAL APPLICATION EXAMPLES

Pennsylvania State University's Distance Learning

Pennsylvania State University in partnership with the Pennsylvania Educational Communications System (PECS) makes televised courses available to more than 700,000 cable subscribers. PECS is a private microwave network built to deliver credit and noncredit courses to subscribers' homes. PECS provides a full-time 24-hour channel through participating cable companies.

This university was the first to experiment with using compressed-video technology to deliver courses. Compressed video maintains a relatively smooth-motion image using a fraction of the bandwidth for traditional television. Interactive compressed video is used to offer courses simultaneously at multiple locations.

The live, two-way compressed video is supplemented by a one-way, full-scan video delivered to all the university's campuses by satellite. The satellite link-up makes it possible to offer all campuses certain courses that until recently were unavailable. Further, Penn State is able to offer a variety of video programming to the campuses and to make its faculty resources accessible to the rest of the world (Augustson, 1988).

Starlink

Starlink is a research network for British astronomers. It provides researchers with interactive computing facilities (hardware and software) including use in image and spectral work. The network helps researchers process data quickly from sources such as satellite astronomy, ground-based telescopes, multiple-dish radio telescope arrays, and automatic scanned photographic plates. According to Quarterman (1990), Starlink has brought together astronomers in the United Kingdom in an integrated community and has vastly increased collaboration and sharing. Starlink has 50 hosts at 19 sites, with 950 users in England, Scotland, Wales, and Northern Ireland. Eighty-seven percent of the users are research astronomers.

Online Computer Library Center

Originally incorporated in 1967, the Online Computer Library Center, Inc. (OCLC) services 6,000 university libraries in the United States, Europe, Japan, and Canada. OCLC holds 15 million catalogue records. Each week members add 24,000 new records and OCLC adds an additional 10,000 from organizations such as the Library of Congress and the British Library. OCLC handles 55,000 requests a week through the interlibrary loan system. The Center also provides catalogue conversion and reference services on computer disks.

Probably the largest national network of this type, OCLC is a prime example of resource sharing over computer networks. OCLC operates over leased lines or dial-up connections; it also provides gateways to other information services. Additionally, OCLC, along with other major centers such as the Research Libraries Group, is involved with a project called Linked Systems Project to enhance the easy exchange of bibliographic records and the sharing of authority data (Arms, 1988).

Collaboratories

Concerned with anticipating ways in which a network will be used, the National Science Foundation is spearheading a project called the "National Collaboratory" (collaboration + laboratory). This is a major, coordinated program leading to an electronic collaboratory, or a "center without walls," in which all researchers at universities, industrial and governmental laboratories, and consortia can participate, regardless of geographic location. Remote interaction is essential because colleagues in an "invisible college" may be widely separated; data may be too vast to store in one location; and some of the most challenging projects may require equipment and facilities which are geographically distant.

Today's scientific challenges focus increasingly on remote phenomena that are difficult to access, distributed across space and time, and are conceptually and computationally complex. The collaboratory requires more than a network itself. It needs a complete infrastructure that provides facilitative software, simulation tools to serve as an analog for a "wet laboratory," remote and interchangeable "smart instruments," digital libraries, and accessible data. The collaboratory will support people-to-people cooperation and collaboration, access to expensive and remote equipment and instruments, as well as links with large databases.

The project, in addition to its obvious merits, will enhance research productivity by opening up an untapped pool of research talent at diverse institutions and also provide information that can be translated into products which will help the country's economy (Wulf, 1989).

There will be a wide variety of services available through the collaboratory, including electronic mail, electronic file transfer, and database access. Higher-level applications (Lederberg and Uncapher, 1989) include:

- digital instrumentation offering real-time control and feedback from remote instruments
- multimedia meetings offering video-conferencing with high-quality audio and shared computational whiteboards
- digital mail providing value-added services such as yellow pages and improved addressing mechanisms
- scientific reference service using human and artificial intelligence
- digital journals, peer review, and a digital library.

Longer-term development would include incorporating such new tools as:

- hypermedia conversation support (video-conferencing supported by hypermedia tools)
- intelligent agents (such as robots) which can search highly distributed libraries, schedule coordinated experiments, and other activities
- interoperable data description (data described so that it is understandable to various analysis systems)
- information fusion (techniques for understanding data from heterogeneous sources)
- smart agents for the design of experiments (tools with an "intelligence" to be a real "assistant" to scientists and engineers)
- smart data gathering (incorporates intelligence into the instruments to allow "self directed" data gathering).

U.S. RESEARCH INITIATIVES

Electronic "Super Highways"

By the close of the 1980s, there were initiatives to further develop U.S. high-performance networks, considered as a kind of "superhighway." One initiative dating back to 1989 was the result of a plan for a Federal High Performance Computing (HPC) Program transmitted to Congress by the Office of Science and Technology Policy in 1989. This initiative envisaged the "wired nation" metaphor where small businesses and homes would share in the capability of high-definition television, supercomputers, vast electronic data banks, and high-performance voice, video, and data communications.

Included in this plan was a proposal for the National Research and Education Network (NREN), a federally coordinated government, industry, and university collaboration "to accelerate the development of high-speed computer networks and to accelerate the rate at which high-performance computing technologies both hardware and software can be developed, commercialized and applied to leading-edge problems of national significance . . . " (D. Allen Bromley in the transmittal letter of the Federal High Performance Computing Program). This "supercomputer" highway would connect government, academia, and industry with a network ultimately capable of transmitting 1,000 times more data per second than current networks.

Among the objectives of the Federal High Performance Computing Program is to speed the flow of inventions from the scientific laboratory to the marketplace. Telecommunications can speed up, intensify, and help clarify aspects of the unfolding project. What networks can do extremely well is organize a vast amount of overwhelming data, provide instantaneous communication with significant parties regardless of their geographic location, and provide access to advanced services such as supercomputing and parallel processing.

The goals of the National Research and Education Network (Executive Office of the President, 1989) are to:

- maintain and extend U.S. leadership in high-performance computing, and encourage U.S. sources of production
- encourage innovation in high-peformance computing technologies by increasing their diffusion and assimilation into the U.S. science and engineering communities
- support U.S. economic competitiveness and productivity through greater utilization of networked high-performance computing in analysis, design, and manufacturing.

The Federal High Performance Computing Program stresses increased cooperation between business, academia, and government in building a network that will serve as a prototype for future commercial networks. In proposing NREN, government sources found that the current national network technology does not adequately support wide-based scientific collaboration or access to unique sources, and often the various separate national networks in the United States stand as barriers to effective high-speed communication. Further, Europe and Japan are moving ahead aggressively in a variety of networking areas, surpassing the current state-of-the-art technology in place in North America.

NREN would be built on the existing infrastructure of long-distance lines and fiber optic cables and would employ new transmission technologies to increase its speed and interconnective ability. NREN's structure, based on the existing informal tier system, would be composed of a federally sponsored "superhighway," providing support for large users and access for every state; a middle tier of regional and state networks with a broadband capability; and a lower level of smaller networks such as local area networks at universities.

Three stages are proposed. In the first stage the existing Internet (T1) trunk lines will be upgraded to 1.5 megabits per second, a project already under way. As a complement to this, the Defense Advanced Research Projects Agency is undertaking a project called Research Internet Gateway to develop "policy-based" routing mechanisms which will allow the inter-connection of these trunks. Additionally, directory services and security mechanisms are being added.

In Stage 2 upgraded service will be delivered to 200–300 research facilities with a shared "backbone" network operating at 45 megabits per second. The ability to share this backbone network will reduce costs and improve service. Once the new research backbone is interconnected with the existing NSFNET backbone to NREN, it is anticipated that every university and major laboratory will be interconnected.

Stage 3 plans are still being developed, but extremely high-speed 1–3 gigabit networks supported by fiber optic trunks will be important. Also targeted are advanced capabilities such as remote interactive graphics, nationwide data files, and high-definition television.

Network Design Projects

The development of advanced networks can also serve the purpose of being an opportunity to study network design, applications, economic impacts, or even user behaviors. As introduced in Chapter 3, one theme of Tatsuno's (1986) book on the Japanese plan to use an advanced fiber network to link its "technopolis" cities is that, in addition to serving their own communications needs, Japanese scientists and marketing organiza-tions could research the very technology they were using.

In the United States, one example comes from New York's Centers for Advanced Technology which were created by the state in 1982 with the objective of improving the interface between basic and applied research. These are cooperative research and development centers that bring together New York State, its universities, and its private industry. Jointly these groups engage in basic and applied research, with the aim of harnessing new technologies for the economic good of the state. Currently, every center receives up to $1 million annually from New York State. These funds are matched by at least an equivalent amount from private industry. One of the units, the Center for Advanced Technology at Polytechnic University, focuses on telecommunications. Some 20 corporations, includ-ing IBM, GTE, AT&T, and NYTel, contribute to Polytechnic, and in 1987 the state legislature authorized continued funding at the $1 million level through the 1995 fiscal year.

Networks in the 1990s

A number of new applications will be available over the 1990s to assist researchers and others with technology transfer. As networks move to broadband and faster speeds, new kinds of services will come available. The network of the future will be digital, integrated (carrying audio, text, and video), and completely interactive.

Also to be anticipated as networks become more intelligent are new kinds of software for analysis, such as an "expert system" which uses facts and rules of judgment, expert decision making, and logic to find lines of reasoning which lead to answers or solutions. An expert system incorporates much more than textbook or manual knowledge; it is able to mimic the basic reasoning process of the human expert upon whom the system is patterned (Feigenbaum and Nii, 1988). In essence, a key function of an expert system is efficient technology transfer: storing the vast knowledge of an expert person, thereby capturing a body of experience and expertise and passing it on to others who may be separated by both time and distance. These systems can be used to diagnose a broad spectrum of conditions ranging from human illness to manufacturing backlog. They also are employed to advise, to plan, and to provide specialized knowledge.

TOWARD GLOBAL COLLABORATION

A final and important topic concerns that step beyond extending the network to researchers in remote locations: encouraging them to use the medium for sharing information, education, experimentation, and most of all, collaboration on new ideas and new products. Empirical evidence has shown us that telecommunications research networks are effective for technology transfer, and in numerous instances, essential. The days of the solitary inventor are numbered. Highly motivated information-seeking users have developed new and imaginative ways of communicating with one another electronically; they manage to do so over thousands of miles with electronic mailing lists and news group bulletins, computer conferences, and information exchanges. Groups such as Starlink, the British astronomy network, and OCEANIC, the ocean research group's database, have united researchers with common interests.

Consider the Swedish participation on NORDUNET. In a period of two years the Swedish infrastructure grew dramatically to include virtually all higher education institutions and many corporations and government entities, spearheaded by an aggressive national policy.

Of particular interest to communication researchers studying technology transfer are theories of collaboration. Why do some individuals, separated by oceans, cultures, and other barriers such as imperfect software and hardware persist in collaborating over the networks? Why do other researchers decline?

Part of the answer lies in the essence of the American tradition for competition which inhibits collaboration. In Europe and Japan collaboration and cooperation on large, expensive, difficult projects is pursued. The thinking behind these projects is that no one entity has the resources in terms of money or manpower to undertake the investigation and development of certain products and processes, largely sophisticated high-tech ventures. To assuage this problem, research groups in universities or in consortia are formed where original research can be conducted and findings can be transferred to the participating organizations.

Most nations of the developed world lead the United States in numbers and kinds of research consortia and cooperative ventures. Only in 1984 did the government of the United States enable exemption from antitrust action for certain corporate operations. This was late in the evolution of the global marketplace.

Successful technology transfer over networks can be greatly enhanced by the participation of strong organizational leaders sharing their vision; by a widening in corporate cultures to embrace innovation and new ideas and to support the incubation of these new processes and inventions; and lastly, by government policy that fosters technology transfer through financial incentives, agency assistance, and other positive methods such as the proposed new "high-performance computer highway" and the national collaboratory.

■   ■   ■

High performance networks offer electronic tools that can span geographic, cultural, and technical distances. These tools hold the potential to accelerate technology transfer and to enhance its efficacy; we have only to look for ways to improve cooperation, to work together to make optimum use of them.

7

·

Making a Residence a Home

You may not always agree with Carol Barger, but you do know that she surely conveys the courage of her convictions. Carol is a consumer advocate, so her first thought about any telecommunications issue is whether the customer is getting a fair break—especially residential consumers, and even more so, those who might be disadvantaged by high prices or substandard quality, namely the poor, the elderly, minorities, or rural dwellers. It is not unusual to see Carol at hearings challenging a $10 billion company's evidence that they are using to support a rate hike. Where a rate case might refer to "residential" services, Carol will be quick to point out that we are talking about "families" and "homes." We are talking about children, spouses, single parents, or any persons that choose to live together in a place called "home"—a place of respite, enjoyment, togetherness, and even "personal turf." And when it comes to telecommunications, we are talking about having the security of a phone, privacy, the ability to talk to loved ones who cannot be with us, or the ability to do business or work from home if we wish. Above all, Carol reminds us that in applications of telecommunications we are ultimately talking about people.

Eventually the intelligent network could bring a wide range of new or enhanced services into the home, including voice, information, video, emergency signaling, and telemetry of utilities. The home could eventually be served by a wireless personal communications network that could "travel" wherever you wish (if you wish). In this chapter, we examine the likely direction of residential services including a close examination of videotex, why it has failed in the past, and what could make it successful in the future.

■ ■ ■

LIVING ON THE INTELLIGENT NETWORK

Information, Entertainment, and Emergency Services

We can expect the intelligent network to provide an increasing number of services for our homes, including for example, entertainment, news, education, security monitoring, telemetering of utilities, medical emergency signaling, and the types of services that you would need to work from your home. Further, the coming portability should allow these services to be easily shifted to anywhere you might wish to call "home," including on the road or at the office. As mentioned in earlier chapters, this portability could be integrated for a cluster of personal services, the "PCN," or personal communications network. Table 7.1 lists examples of these services, each of which exists as of this writing, either in demonstration or commercial form, somewhere in the world.

Anything now viewed on television—over-the-air, cable, or cassette programs—could be received as a "dial-up" service. The range of programs would be limited only by the library of the service provider rather than the physical capacity of the telecommunications link. One link into the home could provide these services simultaneous with voice, data, text, security, emergency, or other services. If the system is so designed, television could be in a high-definition format.

Traditional voice telephone will still be important on the intelligent network, but audio fidelity will be enhanced and additional services like

Table *7.1.* New Home Services

—dial-up television services
—integrated emergency alarms (fire, police, medical)
—personal, "transportable" telephone numbers
—personal communications networks (PCNs)
—simultaneous voice, data, or FAX calls
—dial-up educational services
—practical videophone
—direct computer links (no modem)
—other multimedia communications services
—customized newspaper or videotex
—improved home shopping and banking
—multimedia catalogs for shopping
—wireless FAX and personal computer links

---

voice mail, video calling, call screening, privacy codes (a call will be put through only if the caller dials an extra digit or two) will be more widely available. Answering machine services can be obtained directly on the network. Simultaneous FAX will be available to those who may wish to augment the information accompanying a voice or video call.

The morning newspaper can be delivered in videotex or FAX format. Currently on the planning boards are customized newspapers, or newsletters where a customer's "interest" profile is used to select stories of specialized interest from the news wire, and then these stories are formatted and delivered electronically to the home.

Working at Home

In the late 1980s, it was estimated that some 25 million Americans worked totally or partly out of their homes. If they are involved in "information work" (e.g., writing, editing, programming, investing, communicating, designing, researching, consulting), they can conduct their work over the intelligent network. Capabilities for transmission of graphics, photographs, or data add to the opportunities for keeping an "office" in the home. The vision here is not to turn the home into an office, or vice versa, but to increase options for selecting whatever services one might need at a given time.

Going to School at Home

For over three decades much attention has been given to the uses of telecommunications and computing to bring educational services into the home. There is simply no reason for all formal education to have to take place in classrooms. Having more educational services in the home could increase parental involvement, a priority of many school administrators. Tutorial programs, regular class materials, tests, announcements, schedules, or nearly anything that can be communicated in voice, image, text, or data format could be made available from the school to the home. Homework could be done "on-line." Parent-teacher conferences could take place using voice or electronic mail. School board meetings or special tax hearings could be conducted over the network via teleconferencing.

FUNCTIONAL USES OF THE TELEPHONE

What We Know about Phone Service

It is one thing to consider all the possible services that can be delivered to the home. But there is still the question of "Who will buy what?" In this section, we consider prior studies that can shed some light on what people seem to feel they are buying when they purchase communications services for the home. Ironically, the telephone has received very little attention from social scientists. Studying the history of telephone use, as well as current changes, can contribute to our understanding of likely uses of new network services. Keller (1977) has postulated that telephone use has two dimensions, the instrumental (getting things done as in ordering an airplane ticket) and the intrinsic (i.e., talking on the telephone for its own sake). The concept of the telephone's intrinsic uses was further borne out in Wurtzel and Turner's (1977) study of New York City residents whose telephone service was temporarily cut off. Contrary to the expectation that other forms of communication would be substituted for the telephone, it was found that certain uses of the telephone, particularly for making social calls, were not substitutable.

An ethnographic study of telephone uses (Phillips, Lum, and Lawrence, 1983) revealed that people tend to differentiate very much between business and pleasure calls. Business-related calls and some types of social calls to certain people in various relationships to the caller are thought to be more appropriate for the workplace, while social calls and some business calls are more appropriate for the home. Thus, use is associated with location and relationship along the business–pleasure dimension.

Lum (1984), in conducting a series of in-depth interviews about phone use with senior citizens in Hawaii, found two major perceived needs dimensions satisfied by telephone use. The "convenience" dimension was divided into the area of social calls and informational calls. The latter includes the needs to "get things done" and to "get information." The former includes social obligation calls (perceived as necessary to uphold the notion of one's social role), and "I care," or extremely affective calls meant to "brighten one's day," say "I love you," or to lift morale.

Another important observation was that respondents did not consider their telephone use similar to the use of any other singular communications medium. Instead, there was far more emphasis upon its content and use as an interdependent part of everyday life. Nevertheless, Lum found that other media were brought up as being important, especially when respondents were asked to imagine what would happen if their phone were taken away. For instance, one person mentioned that without the telephone the world would "go slower" since there would be no direct contact. Television would be watched but could not substitute for telephone use.

Dimensions of Usage

Another slant is to inquire what broad "functions" customers attribute to their uses of the telephone. This approach is illustrated in a series of "pilot" studies conducted by the author and his colleagues (Williams, Dordick, and Contractor, 1986) in Los Angeles in the year after the divestiture of AT&T. The purpose was to try to determine how people were thinking about their uses of the telephone, now that service was becoming more of a "purchase" than simply paying for a utility.

The largest of these studies was done with college students who, although surely not representative of the larger population, did consistently reveal three main dimensions of usage, which we usually found represented in smaller focus groups of nonstudents. Illustrated in Figure 7.1, these included:

Social: for example, talking with friends, maintaining contact with family, staying in touch

Transactional: getting information, buying something by phone, checking schedules

Emergency: calling the police, getting medical assistance, reporting a fire, delivering bad news quickly

Although these findings remain tentative, they prompt ideas about uses of the network. First, for example, although the three dimensions typically emerged in analysis of the responses of different groups, there was interpretable evidence of how groups might place a different value on these uses. For example, university students placed far less importance on the telephone for emergency services than did a group of retired persons.

Second, to a telephone company, there were implications as to different demands on the network. For example, social uses likely require large blocks of network time; also a customer might be tolerant if the call did not go through (e.g., busy signal) the first time. On the other hand, you would want an emergency call to get through on the first dial, but the network time would be minimal if at all. Just having the service may make one feel more secure. Transactional uses of the residential telephone require only brief amounts of time, but a person is probably not very tolerant about having to make repeated calling attempts as compared with a purely social call.

As new services develop, some thought should be given to the above needs. For example, emergency uses might be easily and more efficiently served by a data rather than voice signal, such as how a 911 call can reveal the home address of a caller. Or when caller identification services become more generally available, emergencies might be signaled by the touch of a single button. Some transaction services could be accomplished by videotex, such as consulting schedules, doing banking, or making purchases. Presumably, when wanted, a video link might be most applicable to social uses of the telephone.

Figure *7.1*. Functional Uses of the Telephone

People's reported usage of the telephone could be classified according to three dimensions: emergency, transaction, and social.

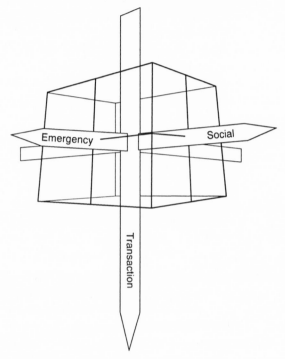

Types of Residential Customers

We can also see implications for purchasing new network services from the examination of differences in current purchases of custom calling and other features. It is well known that there is a part of the general population that say all they want is a simple, inexpensive, reliable voice service, namely, one phone, one line, and a small telephone bill. In the trade, this is called "POTS," for "plain old telephone service." Despite all the fanfare over what is to come in the telecommunications business, there will always be a place for the most basic of services—that is, simply being able to stay in touch with others, and to be able to afford it. On the other hand, if you are enamored of the new alternatives, you might be called a "PANS" customer, for "pretty advanced new stuff." Following is one example (Williams, Sawhney, and Brackenridge, 1990) of how residential customers were divided into three groups according to the types of services desired.

"POTS." Some customers truly want only the most basic of services; currently, this is basic "dial tone" or that they can receive and initiate

voice calls. They have no immediate need or interest in anything other than maintaining basic telephone service and costs at the present level. Most of these customers take their telephone service for granted, but might be price-conscious over a long-distance bill, or if they are in a rural area they might be disturbed that calling to a nearby community involves toll charges. They are usually not interested in hearing advertising pitches for new services. Many do not want to be bothered by choice of long-distance carriers, unbundled bills, and other complexities arising out of the divestiture. They are happy with the basic service and want to be left alone.

"POTS Plus." These residential customers usually have some interest in or experience with something beyond basic service. They are usually aware of their need for extra equipment or services, or the need was aroused when such had been offered. The POTS Plus customer may have an extra line for teenagers or in-laws. They typically know about custom calling features (abbreviated dialing, call forwarding, etc.) and have often made a pro or con decision about buying an answering machine. A few may wish to use a computer modem with their telephone for household purposes or so that a youngster can go on-line. Many of these customers may have reconfigured their phone services when moving to a new residence or when others joined the household unit (youngster returning from school, parent moving in). This group should have an interest in any intelligent network service that would facilitate or improve residential life.

"POTS Plus Business." Residential customers in this category are distinguished by doing part or all of their business from home. Their need for extra services involves small business applications, including FAX, custom calling features, wide area dialing, a foreign exchange line, special long-distance services, and in some cases getting a personal computer on line via a modem. They may also purchase cellular service for business purposes. These customers are a likely market for intelligent network small business services (see Chapter 5), perhaps not for a fully outfitted "office" but for the particular tasks they need to accomplish—e.g., trading on the stock market, editing copy on-line, using graphics systems for design purposes, or doing research from databases.

BUYING MEDIA SATISFACTION

"Uses and Gratifications" View

Another dimension of residential usage might be more identified with media than telephone services. But if the network may eventually carry text services or video, what do we know about people's reasons for using these? Some of the research into "uses and gratifications" theory can help on this point, as it attempts to explain what types of needs are satisfied by people's uses of different media. Chief among these, as originally posed by Elihu Katz and his colleagues (Katz, Gurevitch, and Hass, 1973), are:

- cognitive needs related to strengthening information, knowledge, or understanding
- affective needs related to strengthening aesthetic, pleasurable, and emotional experience
- integrative needs related to strengthening credibility, confidence, stability, and status; these combine both cognitive and affective elements
- social integration needs related to strengthening contact with family, friends, and the world
- tension release needs related to escape, which can be seen in terms of the weakening of contact with self and one's social roles

We next examine a few examples which reflect a uses and gratifications approach in uses of video and information services.

Examples of Satisfactions

In the cable area, one study that deals directly with expectations points out the requirements for researchers to expand their concept of what types of specific gratifications these may include. Shaver (1983) conducted focus group interviews and found functions relating to the unique aspects of the medium. The two most frequently mentioned motives for viewing cable TV were "variety" provided by the increased number of channels and programming choices and "control over viewing" associated with the flexibility of programming. Although the study does not group dimensions in this manner, the motives fall into content (religious programs), structural (variety), service (better reception), and psychological needs (companionship) categories. Some of these gratifications have also been associated with traditional broadcast television (such as general surveillance), but others are especially relevant to cable television.

In a study conducted for the National Cable Television Association (NCTA) (Opinions Research Corporation, 1983), people who choose whether or not to subscribe to cable TV were loosely classified into different user groups. Although better reception was again mentioned as a primary reason for subscribing to cable, certain other distinctions between users were defined. Three user groups reflected the following characteristics:

- "undifferentiated users" of the medium who still watch everything and are early adopters of cable

- "entertain me users" who primarily seek entertainment and diversion and are likely to be pay-TV subscribers
- "basic but . . . group" who are more differentiating in their use; specifically, they want sophisticated, intellectually stimulating, children's, family-oriented, or information programming, and are likely to seek out home services and other special services.

The foregoing study shows that functions relating both to the form and to the content of the medium will be associated with a technology. Furthermore, a distinction seem to be evident between people who seek out entertainment or escape and those who seek a variety of specific services, a point relevant to uses of on-line information, emergency, or "text" services. One interesting uses and gratification study in the information services area by Dozier and Ledingham (1982) identified two key dimensions in the use of interactive banking, shopping, and home security. These were "surveillance" and "transaction." The surveillance mode, a "read-only" status, was found to be more attractice to users than the transaction, "read-write" interactive mode. The researchers speculated that the surveillance function of information utilities will be adopted more rapidly than the transaction one.

Information retrieval systems are sometimes associated with a "depersonalization" factor. Take, for example, systems used mainly for electronic banking, computerized shopping, and home security. The same Dozier and Ledingham study revealed that although there are perceived advantages such as time savings and convenience, such services are often viewed as not being worth the disadvantage of the loss of social interaction. Thus, while the desire for convenience is served, the need for getting out of the house and engaging in daily social interactions is not met for some users.

Further thoughts on uses and gratifications can be found in a book of research papers edited by Rosengren, Wenner, and Palmgreen (1985).

PROJECTION: PROSPECTS FOR HOME INFORMATION SERVICES

The Business of "Videotex"

"Videotex," or the selling of network information and text communication services, has been the most major venture in expanding telephone network applications other than voice. To date, it has largely failed in the United States. These experiences may serve as the best basis yet to consider the future of network markets.

Text information services for the public have typically been classified as "teletext" or "videotex" (sometimes "videotext"). Teletext is usually a broadcast technology where, under one configuration, data are broadcast in the blank between images in television broadcasting. The received information is stored in a memory to be added to a television set or video monitor. Micro-circuitry is then used to decode "pages" of this information so as to display it on the monitor screen. These pages are automatically updated in memory at assigned broadcast sequence intervals. Though this would appear to be an interactive technology to the user, the interactivity at any given time extends only to the capacity of what is stored in the receiver memory.

Videotex, by contrast, is an "on-line" technology, which is an extension of time-sharing networks with large computers. This requires a direct interactive link between the receiver apparatus and the host computer, usually provided by access via computer modem to the public telephone network. As these applications have evolved, one large computer has served as a "host" or "gateway" to other computers from which databases can be accessed. The gateway serves as a main interface for the user which functions to solve protocol problems with other computers, to maintain a "menu" to various services, to store and report billing information, and to serve as a repository in some cases for electronic mail. One of the first popular examples of this system was "Prestel" (originally called "Viewdata"), which was brought on-line in the late 1970s by the British Post Office (Borders, 1980).

Probably the best-known videotex service has been "Minitel" which came on-line for French users in the 1980s (Epstein, 1986). Perhaps gaining the most visibility for their limited life in the United States were videotex trials conducted in the 1980s by large newspaper firms such as Knight-Ridder's "Viewtron" in Florida or Times Mirror's "Gateway" in Orange County, California (Crook, 1983). Both services have been discontinued.

There have also been on-line information services of a smaller scope, and mostly intended for personal computer users. Probably most popular with personal computer enthusiasts in the United States have been The Source, which used *Reader's Digest* computers, and CompuServe which grew out of nighttime use of H & R Block computers (see Magid, 1989; and Aumente, 1987). CompuServe subsequently bought out The Source. To provide an idea of the service alternatives, a generalized "top"menu for such a service is displayed in Table 7.2. An example of a business-oriented, public, on-line service is Dow Jones News Retrieval, used mainly by business customers. Many other services can be accessed by the public via

Table 7.2. "Top" Menu

(This menu is generalized from several services.)

* HELP!

\# Find a Topic

  1. Banking

  2. Bulletin Boards

  3. Chat

  4. Current News

  5. Education

  6. E-mail

  7. Games

  8. Gateway to other Services

  9. Interest Groups

10. Investing

11. News Archives

12. Reference

13. Shopping

14. Travel Schedules and Booking

Enter choice here:

---

modems and personal computers, including private "bulletin boards," but most commercial services have either been short-lived, or had small or specialized audiences. (We are omitting from discussion data-based "references" like Dialog because they are not meant, at least currently, for a large public audience.) A current service to gain some attention is Prodigy, offered in a partnership between IBM and Sears, Roebuck (an earlier partner, CBS, dropped out; see Crenshaw, 1989).

Although CompuServe has continued to grow and Prodigy offers new promises, neither may be even close to a mass market or highly "public" network information service, which they would definitely like to grow into. Why have such services not developed for mass market use? Let us examine some of the likely reasons for failure.

Reasons for Failure

In retrospect, there appear to be four reasons why videotex has not grown in the United States (more details are given in Williams, 1990).

- services offered over the network have been of limited appeal or were available more easily or less expensively in traditional forms (mostly print)

- service providers did not want to be forced into businesses not traditional or not profitable to them (e.g., newspaper publishers are not interested in entering the e-mail business)
- user terminal equipment, software, and hookups were cumbersome, sometimes expensive, and not always easy to use (services seem very slow compared to other computer use)
- telephone companies, a very likely information provider, have been hindered from getting into the business. Regulatory restraint has discouraged entry, thus keeping companies with combined assets of over $100 billion out of the business.

In all, the many failures would suggest that network service providers have not settled upon a needed, easily accessed, and affordable mix of services that would be competitive with nonnetwork alteratives. There is a need to identify at least one highly attractive ("trigger") service that would gain and maintain subscribers (Greenberger, 1985).

Telco Entry into the Business

It appears as if regulatory barriers will become less of a problem for telco entry into videotex. AT&T is now able to make its seventh-year postdivestiture entrance into information services. Although the Bell operating companies are still barred from owning or originating information for videotex services, they can provide the "gateway" function. Moreover, as of this writing, their prohibition from being information providers is coming under close review by an appeals court.

It is important, however, to note an important difference in how telephone service areas are divided. Most concentrations of services are operated by local exchange companies under monopoly "franchises" in "local access and transport areas" (LATAs), defined during divestiture as the noncompetitive sector of the business. This is where the divested Bell companies, large independents (e.g., GTE, United Telephone System), and many very small cooperatives or independents operate. To offer a wide variety of services, a videotex provider must draw from different information providers, many of whom may be outside of a local exchange telecomunications area. Thus, there are costs of importing services which may also involve services of a long-distance carrier. Further, if customers are not given a "local" number to call for their services, there will be intra-LATA tolls involved. In short, any business done outside of a local calling area will place special cost considerations on a local exchange company.

Long-distance companies, typically called "interexchange" companies in the business (e.g., AT&T, MCI, US Sprint, etc.), unless they are connected directly to a customer (a form called "bypass"), must do business through a local exchange carrier for which "access" fees are charged. Although they could operate with toll-free 800 numbers for their information gateways, there could be local access fees.

Either type of telephone company would like to do more in this business than simply be a transport medium or a gateway. They would like to own or have majority control over the information they provide. This is not only the more lucrative side of the business (including higher-priced, premium services), but it also gives them more veto power over what services they vend. If in the business solely as a "common carrier," they are likely to have to transport any information that might have market value (with only very few exceptions, the telephone company does not control the content of your voice calls). However, as a "publisher," the company could control its content much as a broadcaster or newspaper publisher now does. Local exchange companies would like to start with their yellow pages services, which also means bringing their advertising business onto the network. This could be the highly attractive, important trigger service needed for success. If all of this sounds like unwanted competition for newspapers, it indeed is, and that is why the American Newspaper Publishers Association has lobbied so intensively to keep telephone companies out of the business.

Southwestern Bell's experience in Houston trials of videotex services reveals the tribulations of dealing with information service vendors. When the trial was completed, it was noted that a great amount of the revenues that flowed into the system flowed right through Southwestern Bell to U.S. Videotel, an information provider, putting the telephone company not only in a minority partner position, but unable to have much control over the information content side of the business. They also became the target of public criticism when there were any objections to the content or the costs of the services.

Some critics would agree, too, that as telephone companies have entered into videotex trials (e.g., U.S. West, Southwestern Bell), they have made the same basic mistake that the newspaper companies have. That is, they are approaching their new business too narrowly, seeing it mainly within the confines of their traditional business. Again, for example, in the Southwestern Bell trial, assumptions were made about what services might best fit the market, and those customers were given easy access to and even free time for these services. But like most of the earlier videotex trials, there was an early rush of people to try different services, but when the novelty wore off they either retreated from the system or to a service that might serve the important "trigger" function. In the meantime, there were nightmares where people ran up high bills playing games, and more or less nonsensical uses of the system gained publicity. Even given the current barriers to their entry into the business, as telephone companies engage in more videotex trials, they could do well to study carefully the failures of others in this business. Some "lessons" from past experiences are summarized next.

Lessons for New Videotex Trials

Six lessons learned from past videotex experiences include the following concerns:

1. *Ease of use* It should be nearly as easy to log on to a videotex system as it is to dial a voice call, and then there should be a wide variety of alternatives for gaining access to desired services. Not only should there be logically explicit menus, so that without excessive directions, or any at all, a user can move logically through a hierarchy of menus; but there should also be shortcuts in the sense that one can move directly to a given service.

We should see, too, the development of "smart" menus which learn a user's preferences over time, and take them into account when that user logs on to the system. For example, if you typically check mail, move to news headlines, then move to certain stock quotes, a smart menu subsequently can make these moves for you, having observed them in earlier log-ons. We will probably see more use of icons and the kind of window-type interfaces that have become popular on personal computers. Included in ease of use should be methods for handling small crises, such as in what to do when lost in a file or bumped out of a service. For most of these crises, if they cannot be solved automatically, there should be easy "single key stroke" remedial procedures to be undertaken by the user. The ease of use provision should also operate so as to minimize not only on-line charges, but charges for premium services. As is seen in some current software, the premium service might provide batches of data for rapid download so as to overcome high cost. The batches, once stored in the user's terminal, could be pursued at a leisurely pace.

2. *Speed* Again, as everyday users of computers become used to much more rapid processing speeds, it will be necessary for on-line services to avoid waiting times, especially when in the midst of processing successive requests from a user. This means better time-sharing capabilities at host computers as well as transmission speeds probably of 9,600 baud and up. (Many people now log on to information services at 300 or 1,200 baud rates.)

3. *Market Flexibility* The British were moving toward market flexibility with Prestel but it has been better achieved by the French use of Minitel. In an approach that the present author likens to a "flea market" strategy, the on-line system market could have the following features. First, everything is done to increase the likely availability of customers for services that are going to be vended. In short, every tactic should be undertaken to make it easy and inexpensive for customers to be on-line and even to try new services. This is not unlike the low cost of coming to a flea market with incentives such as free lunches, refreshments, and the like to lure customers. Second, it should be easy for vendors to offer their services, just as usually for a modest fee a vendor can have a booth in a flea market. This will encourage the entrance of small, innovative vendors to try their wares in the public electronic market. Third, the customer as well as vendor system should offer easy provisions of exit as well as entry. The goal is to "find the market" so to speak, so that there is an evolution of vendors and customers to support a financially successful service. The publishers, in particular, had incorrect and often inflexible attitudes about the videotex market.

4. *Simple, Inexpensive, and Easy-to-Use Terminals* As with Minitel, the simpler type of terminal will likely prevail. This eventually can be mass manufactured where the price could come down to below $100, with the customer having a choice of payment over time or a lease option. There could also be some incentives for service providers to subsidize the cost of terminals. If terminal emulation is to be done via personal computers, this software should be designed to have all of the ease of use of the simple terminal, perhaps even a very similar if not identical interface. As already suggested, the usage should proceed at a higher rate than the usual 1,200 baud rate.

5. *Advertising* If one is to envisage the growth of videotex for mass usage, it is likely that a source of revenue beyond subscriber charges will be necessary, and this will be from advertising. There are many ways that advertising could take on new forms in videotex transmissions, many of which have either not been tried or tried on too modest a scale. For example, advertising could be developed that is much more like a point-of-sale presentation than a more passive description of the product. Like the amazing growth in 800 number selling, videotex advertising, which offers the customer a chance to make the transaction immediately, will probably be attractive and could be one of the most efficient methods for sales. It is also possible to conceive of advertisers offering "free screens" where an advertiser's message may appear in a window or in a screen frame, while the service being used is an open window. In this way, the advertiser could subsidize the user's on-line fees, not unlike the way advertisers subsidize the

user's cost of a newspaper, magazine, or television program. New strategies for advertising can be developed exclusively for videotex which make much better and imaginative use of the medium than the few examples that we have seen to date. Prodigy is now moving in this direction.

6. *Innovative Services* One of the generalizations in the historical study of communication is that when a new medium is introduced, it often looks very much like a traditional version. Thus, Gutenberg designed his type to look like writing, the first theatrical films looked like movies of stage scenes, and the first video games had a similarity to their pinball counterparts. It is quite likely that as the market and experience grow in videotex services, new and innovative applications will come along. For example, in most videotex trials it has usually been found that there is a segment of the market that very much wants to use the medium for social purposes, meeting people through interest profiles, exchanging views in forums on esoteric topics, and just generally having a wide range of communication alternatives. If the systems are sufficiently flexible, users as well as vendors will discover innovative applications of videotex and the market will grow in that direction.

In all, it seems inevitable that the right videotex formula will be found in this country and it will join the ranks of the popular media. It may be at the right place at the right time with the coming of the intelligent network.

■ ■ ■

Eventually, residential services could easily be the largest growth market for telecommunications products and services. But most important is how these innovations contribute to the quality of home life, especially in its most personal and humane aspects.

8

---■---

Bringing Innovation
to Public Services

Bob Balentine, superintendent of Park Ridge Schools in northern Bergen County, New Jersey, has found his recent experiences with the new interactive television system so sufficiently exciting and rewarding that he hopes to devote time to writing on the topic upon his retirement next year. A few years back, he and a group of fellow superintendents were faced with declining enrollments and the financial problems this brings in maintaining a full curriculum. With considerable help from the County Vocational Schools director and encouragement from New Jersey Bell, the group decided to install a fiber-based network between their schools so they could share the services of some of their best instructors. Unlike typically rural applications of distance learning, this suburban application was meant to overcome "distance" in the form of traffic congestion and travel times if teachers physically were to go from school to school. In the fall of 1990 Bob's classroom went on-line, hosting a precollege reading course with another school, and participating in two sections of Latin and one of business math with two other schools. "The system not only lived up to our expectations," reports Bob, "but we already see many advantages. We see new ways to open the curriculum in our schools, including teaching subjects (e.g., Japanese language) that we could not afford on our own. The system saves dollars by spreading the services of the best or the most specialized teachers over multiple classrooms. Teachers can videotape sessions to use for makeup by absent students. And we are finding that kids from different districts enjoy getting to know one another." Bringing up a system is not all that easy, cautions Bob, who cites problems with school differences in vacation dates and class schedules, getting term papers or tests distributed among schools, obtaining funding for a new project in an era of increasingly tight budgets, and working around a recalcitrant state education office. "But we have it up and it works," declares Bob, who sees an important future for interactive television in suburban and inner-city schools.

Distance learning is but one example of how telecommunications innovations can contribute to enhancement and productivity in the delivery of public services. The problem is not so much the absence of the technology as it is one of diffusion of innovations into traditionalistic and bureaucratic settings. In this chapter, we examine uses of telecommunications in education, health care delivery, and a few additional public settings.

■　　■　　■

NETWORKING IN THE DELIVERY OF HUMAN SERVICES

Just as we illustrated in Chapters 2 and 4, networks can extend the scope of service delivery and management not only in business but in public service applications. In the former case, where what is delivered is in "information" form, the network can directly serve the client. Examples of this are "dial-in" information services or "hot lines" (often involving 800 toll-free numbers), "distance education," as in the opening anecdote of this chapter, where instruction is shared among schools or directly to the home, and such point-of-contract services as third-party billing where a client's "card" is presented and verified, and the costs communicated on-line by a physician's office, dentist, or pharmacy. There are currently experiments to extend such point-of-contact links to supermarket checkouts so food stamp purchases can be registered electronically.

Public service management via telecommunications is very much like the business examples described in earlier chapters. Here, management is from a central office perhaps overseeing a system of widely dispersed branch offices as in a state employment agency network, health clinics, or tax offices. The advantages to management—e.g., large coverage, rapid information dissemination, instant feedback, access to files, and electronic mail—are largely the same as in business applications. Managers can manage more, costs are less, and if all is in order, productivity increases. Only in this case the benefit is lower costs of service delivery, being able to deliver more services for a fixed amount of funds, or more rapid or effective services, not profit.

Figure *8.1.* Public Service Network Configurations

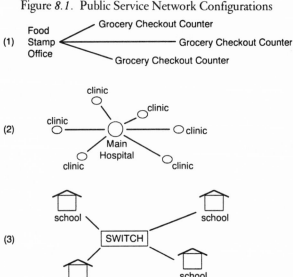

Figure 8.1 summarizes configurations of the major different types of public service applications, namely (1) point of contact, (2) hub, and (3) networking.

A BIRD'S EYE VIEW OF APPLICATIONS

One of the traditional problems in the delivery of public services is the lack of proximity between the client and the service provider. Sometimes this is sheer distance (as in getting medical service to rural areas), or it can be traffic congestion and tight schedules, while other times it is social, cultural, organizational, or bureaucratic barriers. Telecommunications has already been applied to many problems in the public service area, as in 911 emergency calling, "hot lines" and 800 toll-free numbers for specialized information needs, distance education via television, and network applications for medical information and administrative systems. The intelligent network offers many new options of delivering increased, and often transactional, services over wide areas and to remote locations. Table 8.1 summarizes examples.

As we have already discussed, one of the most promising, yet least-used, applications of telecommunications is in delivering educational services. For example, costs of transportation and new school buildings can be reduced

Table *8.1.* Applications in Human and Public Services

—Distance learning
—On-line consultation for physicians
—Coordination of medical service agencies
—On-line consultation for patients
—Multimedia transmission of law enforcement materials (fingerprints, "mug shots," photographs of evidence)
—Expanded 911 emergency services to paging and other portable access
—Job listings and on-line interviewing
—Point-of-sale electronic good stamps
—Public access to health information
—Disaster relief warnings and hot lines
—Resource conservation systems

through shifting more instruction to the home environment. The intelligent network, coupled with the new generation of personal workstations, will enlarge capabilities for these applications. Distance learning also provides important opportunities for continuing education in professions where constant updating is taking place (e.g., as with scientists, engineers, physicians, professors). The network can also support delivery of instruction to rural schools that might not otherwise be available due to distance, personnel, or budget limitations.

Medicine can also benefit. Communication of diagnostic information is a high priority in applications of the intelligent network for medical applications. Consider, for example, a clinic physician sharing a high-definition X-ray diagnostic image over a network where it is possible to enlarge sections of the display, point to selected features, and discuss the case with consulting physicians at a major hospital. Administration is also a major application where, for example, emergency communications, diagnostic dialogs, and billing can be coordinated among cooperating institutions over a multimedia network. Database access continues to be a priority. MEDLARS, a medical database system for physicians and researchers, was one of the earliest large-scale practical uses of network information services. Imagine this text-based database expanded to include static as well as moving images, voice files, and extensive search capabilities in globally located databases.

There were several large-scale field experiments in the 1970s (see Hudson, 1990) on the use of satellite communications to link paramedics at remote sites with collaborating physicians in central hospitals. In recent years, the use of telecommunications to coordinate medical evacuation helicopters has become an important service as rural medical facilities have diminished in the United States. Many "single office" physicians use telecommunications links for central analysis of electrocardiogram readings. Having a medical alert "button" in the home is a growing reality.

Law enforcement agencies are typically the largest users of telecommunications services in local and regional government. This ranges from links between field units (e.g., patrol cars) to networks connecting different agencies. Increasingly, these are becoming advanced, digital networks, where, for example, a "run" on an auto license plate number can be done from a portable keyboard rather than through error-prone voice transmissions. Fingerprint identification can be done over a network having access to a national database. Home security monitoring can be linked directly to security or police offices. In disasters, medical emergencies, or production stoppages, immediate access to information and messaging systems greatly facilitates gaining the required services. Already well known to many U.S. citizens are the emergency "911" services whereby a simple dial-in will inform authorities of an emergency.

Matching prospective employees to jobs is a "natural" as an intelligent network service. Database capabilities provide rapid access to personnel or job listings. Interactive video links can be the basis for remote interviewing. A plan proposed in Los Angeles, but not implemented due to early telecommunications restrictions in the 1970s, was to install local "interview" stations in public libraries throughout the city. With such facilities, job candidates would need to go no further than their local library for job information and preliminary interviews.

The "point-of-contact" type of service mentioned earlier can extend to the use of debit-type cards as a substitute for food stamps. Just as a point-of-sale transaction might be entered into the network via a computerized cash register, so can an entry be made for food assistance.

Gaining access to governmental information is often a time-consuming, laborious process, even if one knows where to start. This might vary from information necessary for filing a building permit to gaining data from a public file on corporate tax reports. In a program recently initiated in Florida (and described later), citizens can directly access public information via the network and links to public databases. Eventually, such systems may allow for two-way transactions, such as applying for the building permit electronically. In some agencies there are public database listings of requests for project bids, as well as listings of awarded contracts.

Not just communication of warnings about impending emergencies (e.g., hurricane, earthquakes, smog alerts, traffic jams), but the coordination of relief efforts is facilitated greatly by advanced and reliable communications. Much attention has been given in recent years to creating backup systems for intelligent networks (the Los Angeles Airport is one such example), where extensive damage to surface or underground networks can be bypassed by broadcast-based telecommunications (microwave, mobile radio, satellite). Coordination among fire, paramedics, and rescue squads can be greatly improved with modern telecommunications.

In other applications, resource conservation can benefit from the use of sensors for detecting forest fires, malfunctioning water systems, or more general emergency situations. Also, network services can assist in the management of natural resource use. One major example is in computer-assisted irrigation systems. Soil sensors register levels of moisture, information from which is fed into a program that has information on the type of crop, fertilization treatments, growth cycle, and weather pattern. Based on this combined information, just enough water is released to meet the plants' needs. Estimates are that up to 50 percent of irrigation water can be saved by such monitoring and controls, although the costs of these systems are high.

THE CASE FOR DISTANCE EDUCATION

As the Name Implies

Distance education is what its name implies, namely delivering instruction beyond a single classroom or school.[1] Some say the term dates back to the 1880s when the Australians had a system of delivering printed school lessons to persons living in the outback. Since the turn of the century most industrialized nations have had one type or another of "correspondence courses." Broadcasting became a high-visibility component of distance education with the development of Britain's Open University in the 1960s, where educational programming was combined with mail-based correspondence courses.

In the United States we have seen many and varied applications of distance education. For example, telephone companies have long made available "home instruction" links whereby bedridden children could interact with their classrooms. On a much larger scale have been development of radio-based networks as in the states of Wisconsin or Texas, a University of the Air operating out of Nebraska, and Instructional Television Fixed Service (ITFS) where lectures are broadcast in a limited-range area and students may interact via a telephone-based audio link. The latter has been used quite successfully in the dissemination of engineering courses to nearby industry, as in programs operated by Stanford University and the University of Southern California. Although there have been several major initiatives to develop an institution in the United States similar in recognition to Britain's Open University, a comparable system has never been accomplished. There are many opinions on this matter, but a common one is that with our extensive system of community colleges and state universities, students have more of an opportunity to go to a nearby campus than do students in Great Britain.

Educational television, however, has achieved wide use in the United States. This was given a boost when many community or university-based educational television facilities joined in the Public Broadcasting System. In the 1960s' War on Poverty, a television program designed to improve school preparation for economically disadvantaged preschoolers went on to become a landmark in television education. This, of course, was the famous "Sesame Street" which brought many innovative production and instructional strategy methods to the small screen. When "Sesame Street" was produced, it was anticipated that it would be viewed mainly in the home rather than school.

Today, two types of educational broadcasting are found in the nation's public schools, with a third now just getting off the ground. The former include offerings disseminated over traditional airways, usually an educational television station, and often available in alternative form on videotape. The second are programs broadcast over special satellite-based channels to schools' receiving dishes, a method usually called "multipoint distribution." (Later in this chapter we review one such service, TI-IN, a commercial service operated out of San Antonio, Texas, which has many public school clients.) Instruction delivered by multipoint distribution can incorporate a feedback audio link.

The third type of service, as described in the opening anecdote of this chapter, involves the linking of schools on a switched network capable of transmitting interactive compressed video. This is like the usual television image except for a slight "wiping" effect in moving images. (Compressed video requires much less bandwidth than commercial television images, hence the slightly reduced quality.) With switched network capabilities, classrooms can "dial in" to join other classrooms, much like a video teleconference, in order to share instruction from a host site. Any classroom in the network can operate as a host. Students can see one another in different classrooms, and teachers can see selected students, which overcomes the barrier placed upon feedback when only audio is used.

TI-IN: A Multipoint Distribution Network

The multipoint distribution architecture of typical satellite-assisted distance learning services is illustrated in Figure 8.2, a "hub" concept in a telecommunications network. Instruction is disseminated from the head-

Figure *8.2*. Schematic of an Instructional Satellite Network

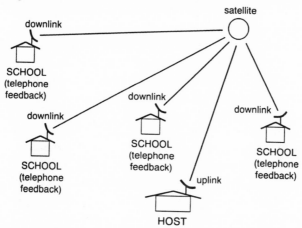

quarters site in San Antonio, Texas, via satellites, to the receiving dishes of schools subscribing to the service and the particular course. The return loop is audio-only, typically via the public telephone network.

If you are a school subscribing to TI-IN, you will need an earth station ("dish"), receiving equipment, cable connections to television sets, and a conference telephone link for audio feedback from classes. Purchasing and installing the necessary equipment may cost between $10,000 and $15,000, and the annual subscription, depending upon the courses used and number of students enrolled, could be in the $5,000 range. You will find the TI-IN people, headquartered in San Antonio, Texas, very knowledgeable about distance learning and highly expert at getting your school started with it. They have been in business since 1984, when TI-IN was started by a private group for the chief purpose of upgrading rural school offerings in Texas. Since that time, TI-IN had grown by 1990 to subscriptions with 950 sites in 30 states.

TI-IN's courses may especially assist a school in offering advanced courses that might not otherwise be available because of lack of teachers or resources, especially in rural areas. This would include advanced courses in science, mathematics, or foreign languages, all important for college entrance. They could also offer a more exotic course not usually found in the local curriculum, for example, Japanese or Chinese language instruction. TI-IN also offers services in teacher, administrator, and school board member training. Many of these latter services are necesesary for state certification requirements, so there is a natural market for them.

TI-IN is also an effective means for dissemination of specialized materials to schools, one newly considered application being instruction about

AIDS. Items on the planning boards include improving means for schools to call in for technical assistance as well as the possibility of an electronic mail network. Interestingly, as compared with many earlier distance learning undertakings, TI-IN is a private enterprise operation which has developed successfully with public institutions as its main customers. It is also a partner in federally subsidized distance education initiatives such as the United Star Schools Network, a subsidiary organization which with a $9.7 million grant serves (as of this writing) 316 schools in 20 states.

Many persons when visualizing the use of a television-based instructional system imagine large groups of students watching a "canned" course, which is not at all the approach of TI-IN. With TI-IN, students enroll in the course and meet in small groups at their school with an in-class facilitator. Lectures are focused on the small group, although there may be many small groups across different schools. There are successful strategies to promote class discussions and activities, and, as mentioned earlier, students can use the phone link for asking questions of the instructor. TI-IN developers deserve considerable credit for avoiding many of the traditional pitfalls of "mass" distributed educational television, which students often find boring. On the contrary, the TI-IN classroom has a distinctive small-group and, as much as is possible, an interactive feeling. Because TI-IN's success or survival is tested in the marketplace, most observers would note that the courses are well prepared and executed. In short, they usually meet the challenge of keeping students of different types and from different regions of the country engaged in the instructional process.

Relative to the challenges of implementing a technological innovation in education, TI-IN deserves special recognition for its success in the diffusion and adoption process. The technological component for this type of distance learning network is fairly common in business and other applications, so the use of satellite is not the exceptional contribution of the TI-IN organization. Schools are typically slow adopters of new technologies; they often have budget limitations for anything new; and many schools, especially rural ones, are wary of "outside" instruction. Moreover, there are teacher certification requirements that vary across states, differences in the requirements specified for given courses, and variations in the degree to which a district has to have county or state sanction to contract for outside educational services. Given these barriers, it is a wonder that a system like TI-IN has done so well.

Beyond the barriers cited above, what are the negative aspects of using a multipoint distribution service? In our experience in visiting user sites, the greatest cited problem, as with most broadcast educational services, is getting the right "fit" with the local curriculum. This is not only in terms of course content and approach, but in terms of something as simple as schedule. A perennial problem is that broadcasters' schedules and schools' class schedules often do not coincide. No matter how much the broadcaster tries to fit school schedules, there are always schools that have conflicts. Although the time problem might seem trivial to those outside of school management, it can create many problems with students' schedules, such as having to miss part of another class, arrange free times to solve conflicts, or have the class at a very early or late hour. If the program is recorded to overcome schedule problems, then the opportunity for live feedback is lost.

Another problem is continuing cost. School budgets are often very tight and always subject to public scrutiny. It is not unusual for a community member or an unsophisticated school board member to relegate distance education to "watching TV," and to question why schools engage in this while parents are trying to cut down television viewing at home. Also, in small rural communities there is the valid argument that they should "keep their money in the community." Why not use distance education funds with other monies to establish a new teacher position, thus creating another job in the community for a person who will spend most of his or her income there? The shortcoming is that although the new teacher might fulfill one need, there will still be other courses that cannot be offered without outside help.

Nonetheless, TI-IN has accomplished what many other projects or systems have failed to do, and there is an important future for it and related distance education services.

FURTHER NOTES ON MULTIPOINT NETWORKS

TI-IN, of course, is not the only example of uses of multipoint networks. There are other applications, some initiated by districts, some by cooperative agreements between districts and community or state colleges, and some where an existing dish—say, from TI-IN—is used for further services. A current valuable source for other applications is the Office of Technology Assessment's *Linking for Learning* (1989).

The Whittle Communications company initiated an opportunity in 1989 to equip a school district with a dish, receiving equipment, and television sets, in return for an obligation to air a morning news program that includes a brief commercial period. The equipment can be used for other purposes, including further programming initiated by the Whittle company. The offer has stirred controversy in the educational community because of the showing of the daily commercial, and the worry in some quarters that children across the country might be confining their news viewing to a single source. Nevertheless, the service has been adopted in some schools and the number is growing.

In one study (Stover and Williams, 1990), the school system in Demopolis, Alabama, was visited at the end of its first spring use of the Whittle service. The district had adopted the service because there were no funds available to expand the TI-IN system they had been given as a part of a University of Alabama project. In particular, they needed the television sets. In installing the Whittle equipment, local school officials were able to get inside wiring contributed by the local telephone company; then, using much of their own labor, they created a network among their classrooms, the receiving dishes, and a small studio with a camcorder and video cassette players. After the first year of the Whittle offerings, they found the commercials unobjectionable, the news worthy of showing, but went on to use their new system for many other applications, including student creation of their own news program. They are also negotiating with the University of Alabama to receive several distance learning programs originating from the university. The district now uses as much distance learning as some schools that have been subsidized to do so and much of their entrance into these uses was due to initial acquisition of TI-IN and Whittle services, both of which were free (again, TI-IN came as a part of a University project).

One final comment on satellite-based systems is that students report that they would like to have more interaction with the teacher and other students. This has been a longtime reaction to instructional television in more traditional forms. But TI-IN, in particular, probably does best relative to this almost generic criticism by keeping classes small, promoting student interaction with the local facilitator, and with the audio feedback capability. There is also a general finding across many distance learning applications that although students express the desire for more interactivity, they typically do just as well on examinations in these courses as compared with those in "live" classrooms.

Minnesota Distance Learning Network

Again, switched network distance learning involves linking schools on a two-way terrestrial network (sometimes including microwave) where schools can join one another in a form best likened to a teleconference. Each school may initiate instruction that is shared by any school that has planned to participate. Unlike the one-way video of most satellite-based systems, teachers can choose to see students at a receiving site just as students can see the teacher. Figure 8.3 presents a schematic of this type of network, which if equipped with cable or optical fiber can accommodate a quite acceptable two-way exchange of full motion video.

Among the growing examples of switched network applications is the Minnesota Distance Learning Network which was started by a group of educators and citizens in Eagle Bend, Minnesota, in 1983 when they were threatened with the school closing because of a lack of resources. At first the system was a one-way microwave link that brought instruction to Eagle Bend, but soon it began to grow to other schools and became interactive when telephone companies joined in the venture and provided fiber transmission capabilities. Schools initiated this system without financial support from the state but with help from the aforementioned telephone companies, cable-TV providers, and private donors.

A factor in the growth of the Minnesota Distance Learning Network was the consensus that they wanted to retain an influence over curriculum materials and to lessen the need to import "foreign" materials into the state. This type of network application, although restricted to regions within the state, also generally avoids schedule conflicts since most Minnesota schools run on similar class timetables, and there is also no problem in certification of out-of-state teachers. As of 1990, the interactive network was employed by 150 of Minnesota's 436 districts. As the system grows, planners would like to not only incorporate the remaining districts but tie in to Minnesota vocational schools, junior colleges, and universities.

If you visit a participating campus, a distance learning classroom will usually be equipped to serve as both a sending and participating site. For participating, there will be three or four television monitors positioned in the front of the class so students can watch the instruction from the originating site, as well as students at selected other sites. Several cameras are positioned to cover the students and possibly the teacher from a participating classroom. If the classroom is an originating site, usually three

Figure *8.3*. Schematic of a Switched Educational Network

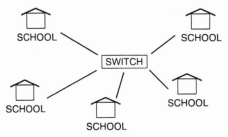

cameras are in use: one to cover the teacher, one to cover the students, and a final one to cover a chalkboard or any other images to be shared.

Users of switched network technology tend to be enthusiastic about their experiences in sharing courses over the network. The two-way video interaction, if well done, tends to support a better feeling of interaction and participation than the one-way video systems. On the other hand, two-way systems require a high degree of host school personnel involvement because they are not purchasing outside course services. They require close cooperation among participating schools. Entry costs are much higher than buying into a service like TI-IN.

In Minnesota, and many other sites wishing to develop a switched network application, there could be an extra "champion" in the picture. This is the local telephone company that, in addition to public service motives, sees development of educational networks as a means to expand or upgrade its network capabilities. Local companies are typically closely regulated monopolies—i.e., public utilities—so unless there is an area of expansion and investment very much in the public interest, it is more difficult for them to expand their business scope. As is recognized in the telecommunications regulatory world, it is difficult for a public utility commission to frown upon upgrades for educational uses.

A major barrier, however, is that under national law, local exchange companies cannot extend their services across the boundaries of certificated areas (LATAs, or "local access and transport areas"). This is the domain of interexchange companies who will object to local exchange companies violating regulations (and competing with them). Unfortunately, many opportunities for distance learning networks involve LATA boundaries. Satellite-based distributors, by contrast, have no such restrictions. Moreover, distance is typically an irrelevant or small factor when participants are within the wide service area of a given satellite. This can be compared with the high costs of running cable or fiber and the need for terrestrial rights of ways.

The Future: Satellite, Switched, or Both?

It is difficult to believe that with the ever-increasing strains as well as disputes over public school financing, that we would not see distance learning promoted as a cost-savings strategy. It is also attractive when a rural school wants to remain accredited (or even open), or when small urban school districts want to avoid being consolidated with their neighbors. So, if distance learning seems inevitable, will it be satellite-based or networked?

This question often leads to short-term controversy because individuals or groups involved in satellite distribution have worked hard to gain acceptance and can easily show that over wide areas, satellite transmission is by far less expensive than installing fiber or leasing existing lines. There are large economies of scale in using satellite to reach widely dispersed school sites. Fiber networking all of them would involve monumental costs.

On the other hand, switched network proponents cite the local advantages of their system, including the highly positive reactions from teachers and students. Naturally, too, telephone companies favor the networks since they may be able to provide them. Satellite services, even if for education, are another case of "bypassing" the local network, a trend which phone companies feel has gotten out of hand, especially considering that they are supposed to be the "certificated" local provider of telecommunications.

The answer to the above question, however, is probably "both." Although satellite has advantages for coverage of large areas, the problem is the high cost of uplink stations if the system is ever to be interactive. Such facilities may be up to ten or more times the cost of a downlink. This becomes an extremely expensive proposition if every district were to make such an investment. On the other hand, it could be possible to include uplinks at points within the larger configuration of networked schools. In brief, these schools could share an uplink and thus realize the benefits of long-haul satellite transmission as well as a degree of full video interactivity. Local or regional distance learning could use the terrestrial switched network, drawing extra resources from commercial hosts like TI-IN over a satellite option if desired. Or when one-way dissemination is the only need, the satellite system and downlinks could handle the traffic without taking up space on the terrestrial network.

HEALTH SERVICES

From Telemedicine to Health Care Administration

The idea of delivering medical advice via telecommunications is not new. It began in the 1920s as radio was used to link public health physicians standing watch at shore stations in order to assist ships at sea that had medical emergencies. Much later came the large-scale demonstrations in telemedicine involving the ATS-6 satellite projects in the 1970s where paramedics in remote Alaskan and Canadian villages were linked with hospitals in distant towns or cities (Hudson, 1990).

Today, the largest applications of telecommunications in medicine are mainly in the administrative area, as in the network applications for health insurance management and third-party billing. This is followed by database services, the best-known of which is MEDLARS, a vast medical reference service. In daily practice, one of the most widespread applications of telecommunications is in the transfer of test information, ranging all the way from EKG input from physicians to diagnostic services, to the dissemination of test results from laboratories to hospitals and physicians. These applications use the public telephone network for either on-line data transmission, or sending of text, or FAX. Among more sophisticated applications are teleconferencing and use of high-definition imagining for transmission of X-rays and other diagnostic images to consulting physicians.

In this section, we examine a few "snapshots" of medical applications of modern telecommunications. Although some may be very similar to business applications—and many are a part of a business operation—such examples are usually grouped in a category of human or public services.

Demopolis, Alabama: A Small Hospital Example

Bryan W. Whitfield Memorial Hospital is a general, short-term acute care hospital located in Demopolis, Alabama. It is the only hospital in the county of Marengo. Its service area extends to a 35-mile radius around Demopolis and also to include all of Marengo county. The 99-bed hospital has 12 physicians on staff and employs 275 people. The hospital has a 24-hour emergency room. Besides medical and surgical beds, it has 12 obstetrical and 7 intensive care beds.[2]

Telecommunications applications in Bryan W. Whitfield Memorial Hospital range as follows. First, the hospital processes insurance claims electronically through computer links with Blue Cross and Blue Shield in Birmingham, where the central processing unit for the insurance claims of the state is located. Two telephone lines are dedicated 24 hours a day to transmit patient data between Demopolis and Birmingham.

The hospital is also networked with two state universities' medical facilities. It sends both its EKG (electrocardiograph) and EEG (electroencephalograph) diagrams to the medical center of University of Alabama at Birmingham to interpreted by the medical experts in that facility. The results come back to the local hospital via FAX. The Demopolis hospital is connected with a dedicated telephone line with the University of Southern Alabama in Mobile for patient care consultation. The University of Southern Alabama Medical Center has a teaching center which is responsible for training and educating physicians in the state.

The hospital is also fully computerized. Patients' personal data as well as medical history are input into the hospital's computer system immediately after they are admitted. The computer also keeps track of the patient's discharge and transfer information. Patient billing, hospital accounting, employee payroll, and medical supply inventory are other functions of the computer.

Since the Demopolis hospital is the only general hospital in the surrounding area, it serves as a referral as well as a consultant center for rural clinics among the neighborhood counties. Seven clinics have direct telephone lines to the hospital. Each only needs to dial an extension number to reach a specific doctor in the hospital.

The hospital purchased its own phone system two years ago to cut its telecommunications cost. It has 25 private lines and an outgoing WATS line, in addition to the hospital's general number. Since the first FAX machine was installed more than a year ago, it has become an indispensable part of the hospital. Other than receiving EKG and EEG interpretations, the FAX machine is used for sending medical supply orders, documents, and letters. Sixty-nine of the hospital's 99 beds are equipped with a television set which receives six different channels from the hospital's private satellite television receiver. The hospital also delivers educational medical programs through this in-house TV system.

Danville, Pennsylvania: Networking Small Clinics

The Geisinger Medical Center in Danville, Pennsylvania, has developed an innovative regional network of outpatient clinics. The system combines the efficiencies of a large referral medical center with locally available clinics to a mostly rural population of nearly 2 million. Telecommunications is a key administrative link in this network. The Geisinger system comprises the Geisinger Foundation and its nine affiliated entities. The foundation is a parent organization to the system's entities and oversees their health-related business activities. The foundation's 14-member board is also involved in initiating and administering grant and philanthropic support for Geisinger units.[3]

Geisinger Medical Center, a nonprofit corporation, owns and operates a 577-bed medical center in Danville. This regional referral center is the system's flagship facility. The Center offers six clinical centers, including the Cancer Center, Children's Hospital Center, Heart Center, Kidney Center, Neuroscience Center, and Trauma Center. The main Center is a Level One Regional Resource Trauma Center and operates a Life Flight rapid response, air ambulance service. All of these centers are served by Geisinger System Services for management and consulting activities. This includes a centralized Information Systems Department that provides telecommunications equipment acquisition and consulting for the system. The main clinic employs a multispecialty physician group, including 457 physicians at 46 sites in 35 communities. Clinic physicians provide skilled services in 65 areas, including acute intervention in myocardial infarction, laser surgery, magnetic resonance imaging, and sleep disorders. The Geisinger Health Plan for this large, mostly rural service area is one of the nation's oldest. It serves somewhat over 78,000 members in 17 Pennsylvania counties.

To support the administrative needs of its 46 sites, Geisinger System Services installed a consolidated billing system. Although individual medical records are kept at each site, the billing system contains basic demographic, insurance, and medical information on patients. Sites are tied to the computer in Danville through dedicated leased lines. The efficiency of the system has developed around a hub-and-spoke concept, with Danville State College and Wilkes-Barre in the hub. Local control is an important aspect of the clinic's success. Doctors are not required to refer their patients to the Center, but usually do. When a local community hospital can provide adequate service, doctors refer patients to that site.

The Geisinger Health Plan is also administered via the systems telecommunications and computer network which handles not only routine record keeping but generates billing. This Plan accounts for around 20 percent of the Geisinger business operation.

Some administrators feel that the EKG reporting system is underused. Physicians in the field have the ability to transmit EKG reports from their clinics to specialists at the Center. However, the system is rarely used because most physicians will send their patients directly to the specialists if there is any suspected problem, whether or not the EKG is sent first.

Another telecommunications-based service offered by Geisinger is the "Lifeline" program, where residents can receive equipment for their homes that automatically contacts the hospital when a special button is pressed. This allows some 100 elderly or disabled patients who live alone or isolated an opportunity to contact the hospital in the case of an emergency.

Finally, there is a Laboratory Information System that is being developed around six hubs. These labs will take specimens, run tests, and feed results into telephone lines connected to the Danville center. Physicians will then receive reports from the on-line system. Currently, seven vans travel 35,000 miles per month transferring specimens and reports. As the tests become fully computerized and reported on the network, this will save costs and time.

In all, the Geisinger system not only illustrates how the different services of a health care delivery system can use common network facilities. It is also an example of new strategies for delivering rural health care, a problem reaching crisis proportions in the United States.

McConnelsville, Ohio: Nursing Home Innovations

The Mark Rest Center, located in McConnelsville, Ohio, and owned by Bethesda Hospital in Zanesville, illustrates everyday uses of telecommunications in a nursing home. It is licensed for 151 beds and currently has 148 residents. Of the occupants, 10 percent are recovering from major accidents or illnesses, while the remaining 90 percent are considered permanent residents. Most of the patients are on Medicaid, Medicare, Worker's Compensation, or other public benefit plans; only 25–30 of the residents pay their own way.[4]

The Center's pharmacy regularly uses FAX for ordering inventory and communicating with local physicians, often solving problems of "telephone tag." The Center's administrative office has computerized room billing, and plans are on the drawing board to computerize the patients' charts. Only 10–15 of the patients have their own telephones; the rest use the facility's telephones in reception areas. In the wing where Alzheimer's patients and others with mental disabilities reside, a wrist alert alarm system has been implemented. The facility's van has a CB radio for emergency purposes. The rest home also has a special cable-TV service.

Several local physicians whose practices include the Center are active proponents of telecommunications to facilitate their work. One regularly uses a modem to access medical information from the American Medical Association and CompuServe databases. The laboratory he uses in Zanesville sends his patients' test results to his FAX early every morning. They see cellular phones in their future as soon as the service is available locally.

OTHER EXAMPLES OF INNOVATION

USDA's Computerized Information Delivery

The U.S. Department of Agriculture has long been an innovator in methods of information dissemination, a key need for rural development. Among these innovations is their Computerized Information Delivery Service which acts as the central node in a computer network that links USDA agencies. The primary object of the service is to make the data in the system available to users as soon as these data are received or at specified release dates set by the agencies. The system, accessible via personal computer modem, serves not only agencies and offices within federal and state government but also private sector groups that will further disseminate the information ("information multipliers"). The system, originally called Electronic Dissemination for Information, became operational in July 1989. The main feature of the system is a comprehensive database of agricultural statistical and narrative data issued by several USDA agencies. The information includes selected reports from the USDA Agricultural Marketing Services, economic outlook and situation reports, foreign agricultural trade leads, export sales reports, world agricultural roundups, soil and water conservation reports, research highlights, consumer information, and press releases. All data on the system are in the public domain.[5]

Florida's Public Access Network and Voice Network

The development of Florida's Department of State Public Access system is one example of new public access opportunities. The system, developed by the Corporate Affairs Division of the Department of State, allows citizens to access the Department of State's corporate records database via remote access. The system went on-line in November of 1987. Previously, information was obtained on a walk-in basis or from telephone operators; but, as the number of inquiries increased, this system became inefficient. A survey taken of telephone users revealed that 85 percent of the calls were going unanswered. The system, which was costing the state half a million dollars to operate, was a source of annoyance for users and managers alike. Realizing that something had to be done to alleviate this problem, Division followed up on a suggestion that would allow users direct access to its corporate records; thus, the idea for the Public Access system was born.[6]

The state of Florida also operates its own voice network, called the SUNCOM Network, with about 135,000 telephones. With leasing and planned equipment upgrades, this network will allow the state to operate and control its own voice and data network while relieving the state of the cost of purchasing the equipment and facilities. The state's telecommunications traffic is estimated to be between 75 percent voice and 25 percent data, with voice increasing at an annual rate of 6 percent and data at an annual rate of 27 percent.

California's Aqueduct Network

An innovative network of fiber optic cables has been installed by MCI for the state along the California Aqueduct system of the Department of Water Resources. The fiber optic cables run north and south across California from east of the San Francisco Bay Area to just east of Los

Angeles. The department started negotiation with MCI by offering them the right-of-way access that the aqueducts provided if MCI would lay down fiber optic cables for department ownership and use. The department wanted fiber optic cables to control its many operations, and MCI wanted to expand its long-distance network. The state's Telecommunications Division has planned to build upon this network by installing fiber optic cables and digital microwave links to connect the north-south network to the main cities of Sacramento, Los Angeles, and San Francisco. In this way, California's major government offices and centers of population will be linked with a cost-effective and state-of-the-art telecommunications network that should significantly improve coordination and provision of state services.[7]

Jackson Center, Ohio: Village Administration

Unlike their urban counterparts, small town, village, or rural county administrators do not have the staff, consultants, or experience to give them the "know how" to best use telecommunications to meet administrative needs. The smallest units often have volunteer administrators who have to tend to other businesses and simply do not have the time to explore innovative solutions to their management activities. As in the city, law enforcement use of telecommunications is often separated from town administration, save for a "hot line" or the village office monitoring police or emergency broadcast channels (from Williams, Sawhney, and Brackenridge, 1990).

Richard Sailor, salaried administrative officer in the village of Jackson Center, Ohio, provides a practical picture of useful telecommunications services. Although his use of voice telephone was fairly routine (which he noted was how he did about 50 percent of his business), Mr. Sailor has explored some uses of telecommunications for special matters (he is an electrician by trade). The village owns its power company, which involves monitoring of operating conditions, including some water abatement problems. He has considered installing a telemetry link to monitor water levels (in case of sump pump failure) and to do the same for some operational facets of the power station. Sailor knows that there are better ways for him to tie into the police and emergency service than to have to monitor their radio communications. He also has been involved in local give-and-take over installing (and paying for) 911 services in the area.

The village has never been especially happy about its cable-TV experiences, although cable is now available and expanding its customer count. Different companies were earlier involved but never built the promised plant. The current company provides relatively satisfactory service but is hard to contact (like some cable companies, they do a poor job of answering their phone). Further, he has a dispute going with them because they are now refusing to include premium channel revenues in their percentage-based franchise fee. In the midst of all this, as well as the above items, Mr. Sailor wondered if the village should build its own cable system and at the same time give itself some village telemetry and emergency communications services, as well as in its school district.

In all, the village would welcome some type of "package" of small-town telecommunications services, not only in the form of equipment, software, telemetry, and network communications, but also cable-TV and database services. Consulting or short courses would also be important.

Lake Milton, Ohio: Land Management

Telecommunications is an essential to modern methods of land management, including the overseeing of park lands. Lake Milton, Ohio, provides an example. Rocco Greco, the Lake Milton Park manager, is a ranger with the Ohio Department of Natural Resources. In his role as park manager, he is responsible for law enforcement and public safety, for the development of natural resources, and for general supervision of all park and lake activities in the Lake Milton Park and Lake properties. In his law enforcement activities, he is connected by telephone and radio to the Youngstown Sheriff's Department; by radio repeater to Columbus and Cleveland; and by computer modem to the Ohio state law enforcement network LEADS (Law Enforcement Automated Data Systems), which tracks drivers' licenses, car registrations, and such. In his water and lake shore management activities, Greco is responsible for a variety of functions, including water levels, so he must know when to open or close the dam. The pool elevation is frequently checked by satellite to ensure the proper lake level. To measure the lake level in a more consistent and accurate manner, the U.S. Army Corps of Engineers is installing a telephone with a voice synthesizer. The synthesizer will be remotely wired so that it can report the levels at any time it is consulted. Further uses for telecommunications are in the planning stage, including an interactive reservations system for lake cabins. (Williams, Sawhney, and Brackenridge, 1990).

■   ■   ■

Public services can benefit greatly from telecommunications applications, but we have been slow to invest in them. Perhaps without competition or vying for the profits, public service agencies, especially public schools, are slow to adopt innovations. Quite possibly, public service budgets will become so burdensome, or quality of services so strained, that pressure will build for investments in productivity. As in the business world, we surely can make some of our public services more efficient through information technology investments.

FOUR

■

Telecommunications and Development

Despite a flurry of studies in the last half decade, the relation of telecommunications to economic development remains an elusive topic for making substantive generalizations. Except for mainly third world projects, we have not systematically gathered the detailed data necessary for a definitive test of the relation. On the other hand, however, we can examine examples where telecommunications has been addressed in the name of development. This is the topic of the next three chapters, each based upon recently completed research projects focusing on state, city, and rural environments.

9

■

States: New Roles and Initiatives

To hear former Senator John DeCamp tell it, deregulating telecom-munications in Nebraska had all the makings of a decisive battle cam-paign. U.S. West and its political supporters, arguing the need to unleash telecommunications in the name of economic development, lined up against Nebraska's Public Service Commission and consumerists, who argued that deregulation would be a disaster for everyday citizens. Sup-porters of Legislative Bill 835 held that opponents were out of touch with economic reality, while the latter protested the "arm twisting" tac-tics of the proponents. In 1986, Nebraska's unicameral legislature passed L.B. 835, at which time the utility commission asked the state's attorney general to file a case testing its constitutionality. The bill was imple-mented in April 1987 and has also been ruled as constitutional. The consequences? Mostly this legislation has gone down in history for all of the debate it caused at the time. Telecommunications and economic de-velopment in Nebraska have turned out neither as a stunning success nor an unfathomable consumer disaster.

The divestiture of AT&T opened new options for states in the regulation of telecommunications. This also has caused them to take a second look at their own use of network services, as well as to consider how telecom-munications can contribute to economic development. In this chapter, we describe several studies into this area and suggest implications for what is yet to come.

■   ■   ■

THE NEW FOCUS ON STATES

Divestiture brought with it an increased focus on state regulation of telecommunications. Essentially, states were left free to lessen regulation of local exchange companies in areas where there were no federal or antitrust (modification of final judgment) prohibitions. They could deregulate AT&T's intrastate service altogether if it could be demonstrated that the company was "nondominate," meaning that its market share was not great enough to control prices.

Upon divestiture in the mid-1980s, local exchange companies in many states moved first to gain price increases, which had mixed results for the most part, although local rates have gone up. AT&T moved to get deregulated, which also had mixed results, although long-distance rates have gone down. In the midst of attention stimulated by these cases, arguments were raised about the importance of telecommunications for economic development. States also began to think more about their own uses of telecommunications, costs as well as modernization, and not only for administrative applications but as an investment in infrastructure that would promote economic development. Telecommunications, which had been a sleepy utility for the most part, began to get some new attention at the state level.

In the early 1990s we have a second round of activity, this time focusing on creating more incentive for the regulated local exchange companies to make infrastructure investments in return for having more flexibility on profits. AT&T still seeks deregulation but may argue more now on economic development terms rather than strictly regulatory flexibility ("freedom to compete") ones. The Nebraska bill may go down in history as one of the opening shots.

During this time, many of the effects of divestiture have evolved as anticipated, at least in terms of introducing competition into the telecommunications business. Local exchange companies have prospered more than most expected in their regulated markets. AT&T has been slower to enjoy regulatory flexibility in intrastate markets (little did they anticipate fighting this battle so much on a state-by-state basis). They are not hurting by any means, except that their entry into the computer market surely was a disappointment. As was the intended consequence of divestiture, the interexchange market has thrived under competition, which has grown steadily, even though AT&T's market share has declined. To look at the business figures, if not the television ad battles, telecommunications is a thriving industry in the 1990s.

A major dividend of divestiture is that states seem to have discovered other new issues related to telecommunications. Certainly, some of this attention has grown out of legislative debate or regulatory hearings where flexibility has been argued in the name of state economic development. Other attention has been raised by the need for large state agencies to make decisions about their own telecommunications uses and facilities, given that deregulation has given them the opportunity to choose among alternative providers. The Bell system is no longer there to provide everything from telephone sets to long-distance services. Certainly, telecommunications has been very much in the business and economic news in the last half decade as our attention has been drawn to its growing importance in government, defense, and especially America's competitiveness in the global market-place. (In the last 20 years, as in too many other industries, we have fallen from being a telecommunications equipment exporter to an importer.)

All of these newly discovered considerations about telecommunications, we sense from our studies of the states, are leading further rounds of regulatory and planning activities. These are ones in which regulation may be less reactive—in turn, more flexible and proactive—and ones where a state's planning and regulatory activities in telecommunications must be more strategically orchestrated if it is to be competitive in today's world. That is, public utility commissions may have to pay more attention to a state's economic goals, and economic development agencies must under-stand more about the changing regulatory environment. In essence, the challenge is to revise our traditional regulatory models for telecommunications in the direction of new competitive paradigms.

A STUDY OF NINE STATES:
ISSUES OF POLICY AND DEVELOPMENT

Plan of the Research

This study (Schmandt, Williams, and Wilson, 1989) grew out of a series of studies in the Lyndon Baines Johnson School of Public Affairs of the growing state policy role of states in the "new federalism" of the Reagan administration. As more responsibility was left to the states in energy, welfare, education, housing, and economic development (rightly or wrongly) in this era, there was the practical question of how states could respond to the challenge. Although telecommunications had been on a somewhat separate trajectory since deregulation had started in the Carter administration, it, too, became one of the new state issues.

The research was initiated in 1987 as a "new state role" piece of research, not necessarily a traditional telecommunications policy or deregulation one, and economic development entered the picture only because it was a growing line of argument for deregulation. This is important to underscore because at the time, most state attention to telecommunications issues was focused on a review of divestiture issues or rate cases. This study, by contrast, picked up on telecommunications as one of the many issues states were grappling with in the Reagan years. Thus it began not with a hypothesis of whether divestiture of AT&T was "good or bad," but what were states doing in this new-found era of telecommunications regulation or planning? Questions were posed in the following four areas:

States as telecommunications policy makers. What appears to be shaping the state's new role in telecommunications regulation?

Telecommunications and economic development. To what degree do states look to telecommunications, and perhaps their regulatory activities, as related to economic development?

Universal service. What type of basic telephone service is defined in different state activities, and are there attempts at changing this definition, perhaps to represent more advanced services?

States as users of telecommunications. Are states now paying more attention to their own use of telecommunications, now that they must go with multiple vendors? Do states see their own uses of telecommunications as related to productivity or economic development?

Most of the responses to the above questions gravitated toward issues involving local exchange companies, although some interexchange topics were considered. A follow-on study, described subsequently, concentrated on the latter.

Altogether, nine states were included in the study. A major criterion for selection was that there was some kind of regulatory activity currently indicative of change in attitudes toward telecommunications. We also wanted to represent each of the seven regional Bell holding companies, some geographic dispersion, and a mix of urban and rural states. Briefly, the states and their features of interest were as follows:

California: Regulators had already begun to examine alternative approaches to rate making, including lifting the ban on competition for intraLATA message toll service. Los Angeles was also the site of several interesting initiatives in using telecommunications for problem solving, in this case to promote "telecommuting" or working at home as an alternative to driving the congested freeways and thus contributing to air pollution.

Florida: This southern state, although having little regulatory activity in telecommunications, represented a striking growth market where one could examine the relation of telecommunications to overall economic growth. There were also several innovative government programs offering information services to the public.

Illinois: This state, well known for its business aggressiveness, housed what was usually considered to be the flagship of the Bell system; Illinois Bell. There were several examples where telecommunications had been seen as important to economic development, including passage of the Universal Telephone Service Production Act of 1985. This act allowed for regulators to relax restrictions on certain telecommunications services where competition was growing.

Nebraska: As mentioned, Nebraska deregulated telecommunications rates and services, the first such undertaking in the country. Arguments favoring this were in part the need to revitalize the Nebraska economy. Nebraska also houses interesting examples of the use of telecommunications for rural development.

New York: One could hardly do a study of telecommunications in the United States and not include New York or more specifically New York City, which most would agree is the telecommunications capital of the world. New York is known for innovative positions on regulation, and the city in particular for the importance of telecommunications to its economy.

Texas: Texas was also undergoing legislative activity in both the local and interexchange areas. The regulatory environment was adversarial and the state had paid little or no attention to the developmental potential of telecommunications.

Vermont: In addition to representing the New England states, Vermont had recently passed landmark legislation involving a degree of deregulation. The Vermont agreement was under examination as a "social contract" that allowed the service providers more latitude and profits so long as they kept rates low. This act was seen in part as an economic initiative.

Virginia: This state had gained attention in regulatory circles because in July 1984 its utility commission had deregulated AT&T. There were also examples in the state where telecommunications had been related to development initiatives.

Washington: In 1985 the state had passed a "regulatory flexibility act" which represented still another approach to deregulation. Washington also houses several very large commercial users of telecommunications, for example, the Boeing company and Weyerhaeuser.

1. States as Policy Makers

As was the presumption in organizing the study, states were definitely taking new positions on telecommunications regulation. Three broad types of action were represented among the nine states.

First there were examples where public utility commissions had themselves undertaken reviews of state policy and began to institute change, as was the case in California, Illinois, and Washington. A second major pattern was where change was mandated by legislative action and generally turned over to utility commissions to implement, as the case in Washington, Texas, and Vermont. The third type of change was unique to Nebraska, where the legislature implemented deregulation independent of the utility commission (which in this case fought the legislation and sought its repeal).

"Regulatory flexibility" rather than "deregulation" was the term most often used to refer to telecommunications policy changes. Mostly, this flexibility initially centered upon reducing barriers or administrative time taken to set tariffs on special services, and to recognize certain areas that were outside the monopoly and were open to nonregulated lines of business (for example, certain business services).

At the interexchange level, the focus was almost exclusively upon defining conditions under which AT&T could be ruled as nondominant in the intrastate long-distance market and thus open to deregulation. Considerable differences evolved in the approach to defining dominance across the different states. Activities during this time reflected beginning discussion on changing the basis for setting local exchange telephone rates. Traditionally, the method had been known as "base rate, rate of return," wherein a telephone company would identify its costs of doing business, then add an allowed profit (say 12 percent) and that would be the basis for setting telephone rates. The move was an attempt to regulate prices rather than profits so that if a telephone company did increase its productivity, it would be allowed to share in those gains (i.e., be able to make profits above a set percentage).

Among the broad generalizations drawn about states as new policy makers in telecommunications was that the process of policy making often took on the character of the state's way of doing business. For example, in

Virginia there was the "gentleman's agreement" or in Vermont the "social contract." Because of the abrasive adversarial environment in the Texas legislature over telecommunications, the activities there became known to the research team as the "Texas shoot-out"! Along that same line, it was noted that where states had a deliberative type of process where facts could be carefully weighed, those states seemed to be making better decisions than ones where changes immediately became legal issues or went directly to the floor for legislative debate. Among the recommendations suggested for states as telecommunications policy makers is that they improve their capabilities for long-range planning and weighing alternatives before attempting legislative action. It was also noted that PUC activities, because of their legal mandate, often took a narrow view of telecommunications regulations, seeming to focus on, say, a specific rate case at the expense of seeing the larger alternatives and motives for setting the rates themselves.

As mentioned earlier, it was clear that states tended to be moving in somewhat different directions in their approach to policy making. On the one hand, that could be viewed as a considerable inefficiency compared to an earlier period when most telecommunications policy making was done at the federal level and in relation to one company. As an anonymous AT&T executive lamented after the company settled their antitrust case with the government, they now were faced with fifty different governments in reviewing their regulatory status in the intrastate market. On the other hand, it was also generalized that allowing the states to move in different directions in a sense made each state a "laboratory" for evaluating new types of policy initiatives. In retrospect, from the perspective of the early nineties, it would appear as if the activities of states in telecommunications policy making had a long-range effect of drawing renewed attention to telecommunications policy making at the federal level, particularly the issue that too much influence in this policy making had been assumed by the court of Judge Greene.

Florida was noteworthy for its lack of attention to policy-making changes and telecommunications. Telecommunications had undergone rapid growth, and it was generalized by the research team that when there are not economic woes to contend with, there would probably be less attention to given telecommunications. By the same token, among the arguments promoting regulatory change in states like Nebraska, Illinois, Washington, and Texas was that telecommunications was vital for "information age" types of businesses so a state should do all it could to promote the business environment for the service providers. Thus, states that had faced some type of economic problem, such as the decline of manufacturing in Illinois, the bleak outlook for agriculture in Nebraska, and the collapse of the oil industry in Texas, created an environment where the very term, "economic development" could draw attention to most any type of regulatory incentive.

New York had a direct study of the importance of telecommunications to retaining large companies (in view of the flight of Fortune 500 companies from the city), as conducted by the state's Department of Economic Development. New York also had a teleport initiative that was tied to economic development purposes. In California, the Public Utility Commission directly considered the importance of telecommunications infrastructure to California's need to stimulate its economy. In Nebraska, the crisis in agribusinesses was a main argument for telecommunications deregulation.

A final point in examining states as telecommunications policy makers is that there is definitely an increase in the stakeholders involved in telecommunications regulation. Whereas in the past, participants in regulatory activities tended to be the telecommunications industry, utility commissions, and public interest groups, now new stakeholders include large telecommunications users, states and cities as large users, and a whole range of consumer groups who fear deregulation could occur and result in steep increases in consumer prices. The increased number of stakeholders has had its effect, too, on increasing the visibility of telecommunications and telecommunications regulatory activities as a statewide issue.

2. Telecommunications and Economic Development

As anyone who has studied the topic well knows, the term "economic development" assumes a variety of meanings in different contexts. Often in the states studied, concerns with economic development were tied to the crisis of the moment.

For example, New York faced the problem of loss of large companies from the metropolitan area, so development was looked at in part as a way of retaining the businesses they did have. Texas had undergone the oil business crisis, so development was often focused upon creating business diversity. For Nebraska, with its farm crisis, development meant not only improving the lot of the farmer, but attracting other types of business (telemarketing). California was experiencing a flattening of its business growth, so development meant the attraction of new types of industries to California, the "high-tech" image of Silicon Valley influencing an image of new businesses. Illinois saw development not just in terms of attracting new businesses but of revitalizing old ones. Vermont and Virginia, somewhat like Nebraska, saw regulatory flexibility as fostering innovation and enhancing business.

Some Bell companies had developed offices to promote economic development and not always centering upon a telecommunications component. For example, New York Telephone created an economic development department in February 1988 as a response to the state's Omnibus Economic Development Act of 1987. Illinois has had a similar office since 1978. In Texas, Southwestern Bell established an economic development office that not only assisted local chambers of commerce in studying development of their areas, but also took an active role in promoting reform of the education system so as to have a better-trained work force.

Finally, there were teleports which presumed to promote economic development. Perhaps best known is New York City's (actually the New York and New Jersey Port Authority) teleport, followed by the Bay Area teleport in California and Houston teleport. However, it should be noted that teleports are not often as much as they are described to be. Typically they are a concentration of send-and-receive satellite antennas with links into local area data and voice networks. It can be mentioned in passing that the Dallas teleport (not studied in this research project) was developed in conjunction with a business real estate area known as Las Colinas in the Dallas area. The area caters to businesses that, among other things, could make use of the teleport facility.

3. Universal Service

In all states studied, "universal service" referred to a customer's ability to obtain voice telephone service at a reasonable rate and level of quality. Despite the wide use of this term in telephone regulation and regulatory debate, there are really no precise and standard definitions of universal service. (It does not appear in the Communications Act of 1934.) As discussed in Chapter 12, the term comes from AT&T early president Theodore Vail's strategy of trading an offer of widely available inexpensive service in return for government protected monopoly. Universal service becomes a topic in regulatory debates mainly in terms of issues where if a telephone company is allowed to raise rates, those higher rates might displace some of the poorer customers off the network, thus an example of "unreasonable" rates. Of course, a counterargument is the existence, in almost all areas, of financial (tele-assistance) programs for individuals who cannot afford basic telephone service ("link up America," "lifeline," etc.)

Universal service is also mentioned in debates regarding the quality of rural service as compared with urban service. In rural areas, the problem of offering service to a remotely located customer, the occasional problem of line quality, the existence of two or multiparty lines, and problems of getting installations or repairs done in a reasonable time are all loosely considered to be part of the universal service requirement. If a state is to offer basic services to its citizens and businesses, is that "universally" available, even in rural areas?

Many of the regulatory flexibility plans allude to universal service either directly or indirectly. For example, Vermont's social contract provides that telephone rates will stay at a level that makes service equally available to all Vermont citizens. The passage in Texas of a local exchange regulatory flexibility bill included the provision of a tele-assistance plan to assure that a person who could not afford telephone service could gain some financial assistance in installation and monthly costs (in this case they had to qualify in the state's welfare system).

Universal service is also a chief argument of consumer organizations or offices of public counsel who look out for the needs of small businesses. Here, such as in California and Texas, arguments against network upgrading suggest that the telephone company's improvement of the network will benefit mainly large users and be built at the cost of the residential or small business customer. Consumerists argue that the deliberate raising of rates will threaten universal service.

In some states, policy debate directly states universal service as the overall goal. This may be often coupled with implementation of a tele-assistance program, for example, the Moore Universal Telephone Act was passed in 1983 in California.

Of course, one of the basic means—cross subsidies—for keeping residential telephone rates low is directly threatened by deregulation. Traditionally in the Bell system, urban and business rates subsidized residential rates and long-distance subsidized local exchange calling. The push toward regulatory flexibility often reflects an attempt to move the prices of telephone services closer to costs, and this, of course, means removing subsidies. One example on the national level has been the implementation of subscriber line charges which require a customer to pay a flat fee for access to long-distance service. There are also access charges that interexchange carriers pay to local exchange companies for the local initiation or termination of long-distance traffic. As one might expect, the long-distance carriers claim that these charges are too high, while the local exchange companies do not want to lower them because it will reduce their revenues, cause them to raise their local prices, and thus incur the wrath of regulators and consumer advocates.

The urban and rural dilemma may be an even more difficult problem to solving subsidies. Many rural companies keep their prices low or are even able to operate because they receive funds from a revenue pool contributed to or by urban companies. If these revenue pools ceased to exist, the rates of rural telephone service would skyrocket, thus creating a major threat to universal service for rural customers. This issue becomes even more contested when propositions are raised for upgrading the rural network so that it may offer the same advanced services as a city dweller might enjoy. The costs of upgrading the rural network, as well as the low rate of return to cover the investment—if it will ever be covered—make it difficult to see how any type of rural services advance could be implemented without preservation of some type of subsidy. While wishing to remain anonymous for obvious reasons, it is not unusual for an urban service provider to point to profits earned by small telephone companies whose existence is guaranteed sometimes almost totally by revenue pool payments. This is sometimes a particularly ticklish issue when comparing rural services provided by a Bell operating company to a rural area as contrasted with a small independent company. The Bell company is often expected to cover more costs itself than a small company that may draw from a revenue pool.

Among the largest issues looming in the future of universal service is the possibility of upgrading its definition from basic voice service to include data services in the form of text and messaging systems or even further "value-added" services such as found on highly advanced networks (computing for example). An interesting example in the possible upgraded definition of universal service was the report of an intelligent network task force in California.

Access to data services is sometimes included in discussions of whether universal service should also include the right to certain types of public information. For example, should a citizen have a right to certain city or federal government information files? Should there be weather service information, legal information, tax information, and the like? It might be, then, that a much upgraded definition of universal service might include certain types of information as well as communication services. Our recent thinking on this topic and a description of the California report are summarized in Chapter 12.

4. States as Telecommunications Users

Among the major consequences of the divestiture of AT&T is that states have had to give more attention to planning and managing their own telecommunications. Whereas before divestiture a state could depend upon the Bell system for telecommunications planning, the postdivestiture era presented challenges of dealing with different vendors, having to do more planning and integration as a large user, and understanding the new alternatives becoming available in an increasingly competitive market. Consequently, most all of the states studied had engaged in some type of telecommunications inventory and planning, some modest and some more ambitious. This planning has been even more challenging because of the need to integrate telecommunications data services with voice as the technologies are increasingly supported by the same infrastructure.

In Texas, for example, an Automated Telecommunications and Information Council was established which examined the long-range needs of the state agencies, including the merger of voice and data services. The studies resulted in the organization of a council that would attempt to coordinate long- and intermediate-range planning, including eventual migration of the network to ISDN standards. Yet in many states, and again Texas is an example, there would be additional planning groups sometimes not coordinating findings with one another. In Texas, there were also several other studies conducted for telecommunications needs of the higher education system, and these were undertaken quite apart from the work of the Automated Telecommunications and Information Council.

Just as university planning, like telecommunications planning, has often been isolated in states, so has planning for the use of telecommunications in computing by law enforcement agencies. There is somewhat of a tradition that state or city planners allow the police department to pursue its own planning and management in telecommunications. Often within that area, law enforcement personnel are particularly concerned not only with the security of their system but also about interfacing it with different levels of law enforcement agencies on regional and federal levels (as with gaining access to FBI data, for example).

In some states telecommunications planning has led to innovations. In California and Texas, for example, project researchers came across plans to increasingly link social service field offices with the central agency and ultimately to bring clients of these services on-line. For example, applications for health care, food stamps, or driver's license could be increasingly done over telecommunications systems from branch offices. There are plans to replace food stamps with debit-type cards where expenditures could be registered at the point of sale. In Nebraska, a center for telecommunications and information was created and eventually led to curriculum development for training telecommunications personnel as offered at Kearney State College (but the center declined with the change of the governorship).

After the period of this study, it has become evident that more states are conducting large-scale examinations of their telecommunications uses, as can be seen in the form of requests for proposals circulated to national research centers in this area. Perhaps one of the most opportune areas for states is to make more substantial applications of telecommunications in education. States like California, Texas, Nebraska, and Vermont have examples of "distance learning" applications (Chapter 8).

Florida, as also described in Chapter 8, has a dial-in database service whereby for a modest fee one can obtain information from various files including building permits, tax information, and various statistics that the state regularly gathers.

Although long-range planning and coordination of telecommunications seem to be a trend for states in their new role of a user, in none of the nine states did we find coordination to cover all agencies, indicating that there remains a "balkanization" of many state's telecommunications infrastructures.

Policy Implications

States with nonadversarial forums for development of telecommunications policies (like Vermont, Virginia, Illinois) seemed to have more of an eye on the larger scale issues than those where disputes were focused on adversarial proceedings on narrow issues (Texas). It would appear that if just some of the funds spent on adversarial-type proceedings in these states were shifted to long-range planning and careful weighing out of policy issues, a state would be ahead. This tied in closely with the recommendation that there be more attention given to understanding what is meant by universal service, including whether as information technologies advanced so should the definition of basic service. Also tied in with the types of policy deliberation procedures is the need to gather data on the consequences of certain policy decisions. There were no examples in the nine states where any type of careful data gathering followed policy implementation.

A second cluster of policy implications reflected economic development issues. One, for example, was that economic development groups should try to educate themselves better on the benefits that telecommunications can offer in industrial recruitment or retention. It seems that much talent and know-how in telecommunications were going unused in the development area because they were deposited in regulatory agencies and only brought to bear on narrow issues under dispute. More attention should be given to surveying the needs of rural users and telecommunications uses for rural development. If states are attempting to develop their rural areas, they have to understand more of the alternatives involved, including whether city users will bear some of the costs.

Finally, although states are becoming much more active in examining their operations as telecommunications planners and users, it is clear that most still lack overall long-range planning and an understanding of how different agency needs might be served by a common advanced network.

INTEREXCHANGE ISSUES

Background

In the context of the aforementioned state study, it was apparent that some of the critical differences between the changing status of local and interexchange companies were being lost in legislative debates. A fundamental point was that divestiture had left regulation of *intrastate interexchange* telecommunications to state utility commissions, whereas *intrastate long-distance* remained under the Federal Communications Commission. There was a sense that the intended effects of divestiture in moving interexchange telecommunications toward a competitive situation were overshadowed by regulatory flexibility and rate case debates mostly of a local exchange nature. As a consequence, a follow-on study was conducted which focused on the states at an interexchange level.

Also on the broader scale, we noted an obvious correspondence between the state as a governance unit and state interexchange telephone networks and services, whether examining those networks in terms of patterns of commerce or for patterns of public services. In the state environment, we can consider cities as concentrations of urban communications served by local exchange companies, as contrasted with the interexchange network that links these centers of commerce as the infrastructure of the state. An important contrast is that, under current regulatory trends, the concentrated centers of basic telecommunications services in the cities are considered as natural monopolies, whereas the interexchange networks are considered to be a competitive market. In essence, therefore, a state can consider its larger telecommunications infrastructure as especially open to development. Regulation of the interexchange market was largely limited to AT&T, which is to be regulated until it can be shown that competition exists from alternative providers.

There are additionally important components of the interexchange business, including special, often enhanced, telecommunications services provided usually to large public or private users as dedicated or private networks. These include the provision of special access services so that high concentrations of traffic may be transmitted directly from the user to the interexchange network. Whether it is in the public interest to bypass the local exchange network in these cases is a matter of continuing debate.

Eight of the nine states studied (California, Florida, Illinois, Nebraska, New York, Texas, Virginia, Washington) were further analysed for the interexchange study. (Vermont, a single LATA state, was omitted.) Some of the major generalizations are summaried next, omitting points already made in the discussion of the earlier state study.[1]

Bypass

In nearly every state, a major concern regarding the future of the public interexchange network was its attractiveness to large users. When the economies of scale allow it, large users tend to move to their own private end-to-end interexchange network. In another version of bypass, companies link directly via private lines from PBXs to the interexchange network, thus bypassing the local exchange network and avoiding expensive access charges. This is not a total loss to the local exchange company because they often lease the private line; that is, they are their own "bypassers."

Clearly, it is the pattern of large users to prefer a private network. Access charges are among the reasons encouraging bypass. Logically, it is in the interest of interexchange carriers to either avoid paying access fees to the local exchange company or to lobby to lessen them. Making interexchange switches especially compatible with modern PBXs is one way to encourage bypass of the local exchange company; increasing the availability of access points ("point of presence" or POPs) to the network is another. As for lobbying, interexchange carriers can argue accurately that large access rates represent a continuing subsidy of the local network by taxing interexchange services, a situation contrary to the move to competition.

Finally, large users may leave a regulated public network out of a preference for private operation if they can afford it. In addition to lower costs, private networks provide flexibility, security, and advanced technologies where and when needed. Large users were quick to tell us that having private control over their services offers several advantages. In a field of software-defined architecture, a user can quickly tailor a system to meet immediate needs rather than wait for a service provider to make changes. Large users may have managers or engineers more experienced than the local service representatives of the interexchange company. Also there is a desire to distance oneself from regulated services where tariff applications may delay changes or lead to the revelation of proprietary information.

Figure *9.1.* Trends in AT&T Market Share
(Share of the interstate market; percentage of "all minutes." Source of data: FCC
NEWS, *September 21, 1988, Number 4608.*

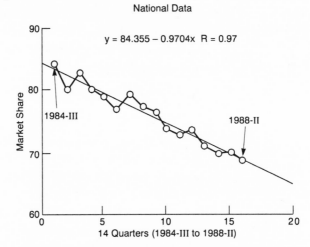

National Data

$y = 84.355 - 0.9704x$ $R = 0.97$

AT&T Market Share

In each of the eight states there were major trends indicating a declining share of the interstate market for AT&T. Again, we use these figures as an indication of choices for toll providers, which is relevant to the intrastate market as well. Figure 9.1 summarizes the aggregate trend when all the states served by AT&T are combined. In this case, the data extend across 16 quarters (1984-III to 1988-II). The correlation between market share as a percentage of all minutes and the 16 successive quarters is 0.97. We must bear in mind, however, that aggregate figures disguise variations in market segments that individually might have quite different patterns, and that regulatory and pricing decisions must reflect these details.

Price Trends

As is well known, the prices of long-distance services has steadily decreased since divestiture. In part, this is a consequence of the new competition, and in part, a reflection of the move to lessen the subsidy of the local network (e.g., by cutting access fees paid to local exchange providers). Figure 9.2 summarizes a consumer price index for telephone service. This is a "market basket"-type figure that represents telephone-related expenditures for interstate as well as intrastate toll calls. It reflects services used by a typical urban household.

Figure *9.2.* Trends in Pricing
(Annual rate of change in price indexes for long-distance service. Source of data: FCC:
Monitoring Reports of CC Docket 87–339; *prepared by the Staff of the Federal-State
Joint Board and CC Docket 80–286, September 1988. The figures for 1988 are
annualized from seven months of data.)*

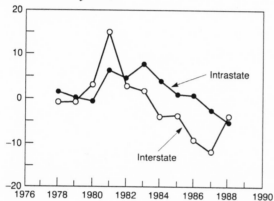

Price and Competition

Still other research bears upon the question of whether price reductions
can be related to increased competition in the toll market. The answer is
yes, at least statistically in an FCC study that calculated the regression of
1987 state interLATA and intraLATA prices on variables reflecting competition, including AT&T's market share. Table 9.1 summarizes the results,
showing that prices can be reliably predicted from competition variables,
although more so for intraLATA than interLATA state markets. Of course,
prediction does not mean cause-and-effect, but we can say that the
postdivestiture experience has seen anticipated price declines along with an
increase in competition within states.

Table *9.1.* Relation of Price and Competition

*(Summary statistics for the regression of 1987 prices on competition. Variables in state in-
terLATA and intraLATA markets; R2 or predicted variance. Source: Chris Frentrup,
"The Effect of Competition and Regulation on AT&T's Intrastate Toll Prices, and of
Competition on Bell Operating Company IntraLATA Toll Prices." Washington, D.C.:
FCC Common Carrier Bureau, June 22, 1988).*

Market	Night-Weekend	Day	Evening
InterLATA	0.779	0.756	0.673
IntraLATA	0.904	0.905	0.905

OVERALL RECOMMENDATIONS AT THE STATE LEVEL

State officials should consider it their responsibility to develop the state telecommunications infrastructure, namely to:

- orchestrate the planning of its different agencies, including public utility commissions and economic development agencies or groups
- plot a longer-range plan for advancing to new network services
- recognize the intended paradigm shift from regulation to competition and differences between local and interexchange companies.
- consider special investments for state innovations
- encourage an infrastructure conducive to meeting the requirements for economic development.

Leaders can begin by developing long-range, user-oriented plans for their states. This plan should be carefully orchestrated with state initiatives for economic development. The state should encourage broad involvement of major service providers in these efforts, including their sponsorship of research and planning activities, which would also involve the sharing of information from other states. In Chapter 12 we provide an example of this in the form of the Intelligent Network Task Force in California.

Generalizations across our state studies are summarized in Table 9.2.

Table *9.2.* Summary of State Study Recommendations

(Adapted from "The Competitive Challenge" [Williams, 1988a].)

The *premises* for state telecommunications regulation should be *regularly reviewed* as to whether, within the range of alternatives allowed by the federal government, regulatory activities are consonant with state goals. Careful examination should be made between regulation meant to protect local monopolies as against that intended to promote competition. "Sunset" reviews of regulatory commissions should be mandated by legislatures.

In order to *avoid the narrow view* in new rounds of regulatory review, the evaluation of regulation should be the topic of high-level, consensus-building state committees or commissions. (The National Governor's Association has developed a policy statement that recognizes the importance of telecommunications in economic development and sets forth network modernization and technological innovation as policy goals.)

States should encourage *research into the consequences of increased flexibility* of regulation. Do the changes lead to the desired results?

If economic development is to be one of the goals of telecommunications regulation, then *a more comprehensive approach should be undertaken in planning*—one that

Table *9.2.* Summary of State Study Recommendations (*con't*)

combines economic planning with regulatory goals, including plans for the state as a user.

Bell operating companies, large independents, and state telephone associations should promote *"stakeholder forums"* so that important users can gain a better knowledge of the applications of the coming intelligent network. The same individuals and groups could, in turn, voice their interests in having modern telecommunications services when state utility commissions are holding hearings on new investment areas. This recommendation reflects the observation that many important citizens are unaware of the benefits of modern telecommunications.

As a basic step in the next round of legislation, national attempts should be made to *define the nature of universal service.* On the one hand, as is often the case, this definition could focus on basic voice service. On the other hand, states may well wish to consider the upgrading of universal services to include the new capabilities of the digital network, as did the California Intelligent Network Task Force (Chapter 12). It is important, too, that access to interexchange services be considered in definitions of universal service.

Assumptions underlying *tele-assistance plans should be reviewed* to determine whether universal service can be accommodated within the financial structure of the telecommunications industry, rather than having to resort to outside programs, especially those of a "welfare" nature.

Bypass should be reviewed in new policy models, rather than treated as an exception to policy, as it seems to be now. Can bypass be incorporated as a positive component as we move to more competitive paradigms for telecommunications policy?

Studies should follow up on *research into relations between competition and pricing,* especially comparisons more of a quasi-experimental nature where more can be said about cause-and-effect. More knowledge on this point is especially important to new initiatives for regulatory flexibility.

The *role of telecommunications in the rural economy and social services* merits special study and incorporation in new state plans and policies. This is also an area of potential state-and-provider partnership in development, as well as an area in which new equipment and systems can likely enjoy a large export market. We need to avoid the "carrier of last resort" mentality.

States without *master plans* for their own telecommunications services should move to create the organizations and processes to develop them. Such undertakings should concentrate on listening to the needs and projections of users. Service and equipment providers should be asked to join the team. The same planning groups should consider the consequences of their plans for regulatory policy as well as state initiatives for economic development.

Table *9.2*. Summary of State Study Recommendations (*con't*)

States that are on the path to upgrading their basic telecommunications services should also consider *strategic investments* in telecommunications that might benefit special industries or developmental plans unique to the state. As in the experience of some states that are already enjoying the benefits of such programs, this is an ideal meeting ground on which business and a state can undertake joint efforts.

Increased cooperation among telecommunications regulators, planners, and developmental agencies should be a high priority in new rounds of telecommunications policy making.

There should be *more research to study the consequence of moves to price caps or banded rates* in place of base rate, rate of return regulation. Is the new incentive truly created, and will the service providers increase efficiencies and, hence, profits? And how will these benefits be shared?

New rounds of regulatory debates should be couched more in terms of the entire U.S. economy. Much of the current debate remains concerned with abstractions about the virtues of the free market and economic efficiency. While important parts of the debate, they may narrow the focus of discussion only to how flexibility will benefit telecommunications companies. A focus on how flexibility may benefit the entire economy is likely to draw the greater attention of regulators and legislators, who are, after all, responsive to the public interest.

■   ■   ■

As telecommunications moves from a regulatory to a competitive paradigm, states should see it in their interest to promote the development of their telecommunications infrastructure. Some are already much further ahead than others and this will likely be reflected in future development on other fronts.

10

■

Cities: New Urban Infrastructures

If you want to do business with our Latin neighbors, you'll want to think about starting with Miami whose business leaders call it the "gateway to the Caribbean and Latin America." This is not just because of the large and active Hispanic population or because almost all Caribbean East Coast cruise ships operate out of their famous port. Mainly it signifies the interational trade emphasis of the Miami business community, and to those in telecommunications it means not only telecommunications equipment for Latin America but first-class business telecommunications links as well. Enrique Lopez, of the management firm Deloitte, Haskins, and Sells, explains how the Miami-based LATCOM, Inc., which promotes telecommunications and information products, convinced the Dade County planning agency ("Beacon Council") and the World Trade Center to put Miami on the international telecommunications map. LATCOM now hosts annual conferences to promote the telecommunications business opportunities in Latin America. Beyond opening those markets, LATCOM's annual conference brings up to 1,000 new visitors and about half a million dollars to Miami. This is a living example of a strategic focus on telecommunications, and it has paid off.[1]

■ ■ ■

WHY STUDY CITIES?

Despite the city being the environment of the greatest concentration of telecommunications infrastructure, there has been relatively less written on policy and planning in this area as compared with, say, national policy or third world examples of telecommunications and economic development.

In terms of applications of telecommunications and industrialized society, Ralph Lee Smith's *The Wired Nation* (1972) reflects upon the importance of telecommunications to modern development. Daniel Bell's *The Coming of Post-Industrial Society* (1976) creates a strong case for information as the key resource in the transition from an economy based mainly upon the manufacturing of goods to one emphasizing services. Bell envisaged "knowledge cities" that would be concentrations of universities, hospitals, research centers, and other information-intensive organizations. Also influential in the 1970s were a series of National Science Foundation sponsored studies into the social uses of interactive cable (Brownstein, 1978), which included planning as well as demonstration projects.

In the 1980s, more was written and discussed in detail from different vantage points about the importance of telecommunications technologies and urban development. Urban researcher Manuel Castells (1985) edited a volume on the spatial characteristics of evolving high-technology societies, including the implications of newly available information technologies. Public administration researcher Mitchell Moss (e.g., 1986), based in New York City, wrote and spoke extensively on the importance of telecommunications to a city's economic development. The Dutton, Blumler, and Kraemer (1986) conference on "wired cities," held in 1984 in the Annenberg Washington Center, brought together individuals working on urban telecommunications at the time. The conference yielded reports on wired city or cable projects in Japan, France, Germany, Britain, and the United States. The conference directors and editors noted, among other generalizations, that U.S. projects reflected more fragmented private planning efforts as compared with national planning in France, Japan, or Germany. Another contributor has been Sharon Strover (1987, 1988) who has pointed out the need to study investments in the telecommunications infrastructure of cities.

Next we turn to a discussion of a study of telecommunications uses and planning in 12 North American cities.

A STUDY OF 12 NORTH AMERICAN CITIES

Background

The urban study described in this section (Schmandt, Williams, Wilson, and Strover, 1990) was partly stimulated by the state study described in Chapter 9. Repeatedly in the study of the nine states, it became clear that the city was an important level for analysis. Paradoxically, the city is the hub of telecommunications concentration, yet cities have very little regulatory power. This raised the issue that in this age of growing global business competition, it is important for cities to understand how the telecommunications infrastructure can contribute to their development, as well as how they might influence policy favorable to that development.

Indeed, in the last several decades cities have had to answer questions about their telecommunications infrastructure, as for example when expertise was needed in planning or reviewing cable-TV franchises, or after divesture when cities found themselves no longer able to call upon the Bell company for total planning. The present study was mainly designed to focus upon a range of American cities representing different economies, different regions, and different Bell holding companies. As the design progressed, two additional North American cities outside the United States were selected for reasons described below.

Cities Selected for Study

There now follows a list of the 12 cities and a point or two on each regarding its selection:

Atlanta: This city was known to be "telecommunications-intensive," including for the reason that it houses a large number of AT&T regional executives, as well as being the headquarters of BellSouth, among the most successful of the Bell holding companies. Atlanta officials, who refer to their city as "the flagship of the South," readily acknowledge the importance of telecommunications for economic development. Indeed, the selection of Atlanta for the 1996 summer Olympic Games has confirmed that attitude.

Boston: The economy of this city had earlier undergone a transformation from older manufacturing industries to growth in high-technology areas ("Route 128") and there was a question of the reliance of these upon the telecommunications infrastructure. Boston represented an example of an older city whose infrastructure had not originally developed with utilities like electricity and telecommunications in mind. As it turned out, much of the modern telecommunications infrastructure of Boston has had to be retrofitted to older rights of way, conduitry, and overhead wires.

El Paso–Ciudad Juárez: Other research by the author and his colleagues (Barrera, 1988; Barrera and Williams, 1990) had indicated the importance of telecommunications for the maquiladora or "twin plant" industries along the U.S.–Mexican border. It has been noted that part of the economic development of El Paso was highly tied to cooperative ventures with manufacturing plants in Ciudad Juárez. Modern telecommunications links were important in many of the larger of these cooperative undertakings, and the issues in transborder telecommunications raised simultaneous U.S. and Mexican policy issues. Ciudad Juárez was one of the North American cities outside of the United States to be studied (Toronto, the other, is discussed below).

Houston: This city is also known as a telecommunications-intensive urban area. One reason for this is that the oil industry requires telecommunications for coordination among its wells, refineries, and markets. Houston was also known to be a city with an aggressive plan for its own city agency telecommunications infrastructure.

Los Angeles: This city had become an important topic in our earlier study of states which had included California. In addition to reflecting aggressive planning in telecommunications, Los Angeles represented several examples where telecommunications had been used in part to solve a pressing local problem. The substitution of telecommunications for commuting is one of the strategies to reduce air pollution.

Miami: We were not only interested in representing this sector of the United States in the study, but Miami (as in the chapter opening) had been developing a reputation as a "gateway" to Latin America for trade cooperation and development. Given the acknowledged importance of competing in the global economy, Miami represented an attempt to be a visible competitor and one which recognized the importance of telecommunications in that effort.

Minneapolis: Other than representing a particular region, Minneapolis was chosen because, among U.S. cities, it had earlier undertaken one of the most comprehensive studies of the likely uses of telecommunications for social services and development. As it turned out, this study did not lead to much innovation on the city's part, but was probably a valuable recognition of how to encourage other interests to contribute to the infrastructure.

New York City: One could scarcely do a telecommunications study of U.S. cities without including New York City, which is acknowledged among most experts not only as having the most sophisticated telecommunications infrastructure, but as visibly acknowledging its importance in economic competitiveness. The New York/New Jersey "teleport" was also a topic of interest.

Phoenix: This is an example of a city outgrowing its infrastructure and one attempting to develop its research and high-technology manufacturing base. It was a valuable addition because of its representation as a key "Sunbelt" example.

Pittsburgh: Although Pittsburgh was not known for telecommunications examples, it was representative of a situation where considerable amounts of economic revitalization had been undertaken to replace the aging steel industry. Researchers were interested in the degree to which telecommunications was a part of this revitalization. Pittsburgh also houses the "wired campus" example at the University of Pittsburgh.

Seattle: This city was selected for regional representation, but also, like Los Angeles and Miami, it was an example of important seaport activity. The port of Seattle offers special telecommunications and data processing services to its clients.

Toronto: This Canadian city was selected for the main purpose of allowing the researchers to compare a major industrial center still operated under a traditional monopoly regulatory environment, to American cities where competition and regulatory flexibility were evolving.

Although the results of this study can be examined on a city-by-city basis (Schmandt et al., 1990), most interesting to us for the present chapter are the generalizations that were developed across the individual cities. These fell into four areas: (1) telecommunications as urban infrastructure, (2) major players in telecommunications growth, (3) the city as a telecommunications planner and user, and (4) telecommunications and urban economic development. Each of these topics is next described in more detail as drawn from the monograph and with occasional further interpretations by the present author.

1. Telecommunications as Urban Infrastructure

Not all city personnel readily recognize telecommunications as a key component of a city's infrastructure, yet most will agree on that point if prompted. It is as if telecommunications has been seen only as a utility where the main goal was to keep costs down. All of the cities in this study had complex and growing telecommunications infrastructures, mixes of voice, data, and video services, networks operating under private and public ownership, and in regulated and unregulated environments.

The idea of telecommunications infrastructure is new to many, but it is a concept that is catching on. Mostly when planners think about infrastructure, they consider roads, water, education systems, or other city facilities. Traditionally, telecommunications has been more or less of a "background" utility. In cities where there is more attention given to telecommunications as infrastructure, it is increasingly seen as an important strategic investment. In most cities studied, a major share of that infrastructure was the result of many years of service by the Bell system, where a city was basically the largest customer, and other more advanced services may have been developed for use by large telecommunications users (hotels, travel agencies, large retailers, insurance companies, and financial institutions).

Although most of the urban infrastructures were wired network (supplemented with microwave and other assisting transmission technologies), some of them were broadcast-based, as in police radio, paging and dispatching networks, or in the near future personal communications networks (PCNs). With the ability to build private networks coming immediately before divestiture and gaining much impetus after, there has been a very visible trend of large users bypassing the public network for their individual needs. This may include networks for voice and data to link the different offices or plants of a company, including local area networks. It also includes the linkage of the telephone systems of large businesses directly to long-distance providers, bypassing the local exchange company.

Some of these large users are bypassing the interexchange network as well, as they directly link their offices nationally and internationally. In various cases, these bypasses are bought by cities themselves as large users, thus adding to the influence of larger users in *de facto* policy making regarding bypass.

The degree to which microwave and eventually satellite are involved in these networks may depend also on a city's geography. Where line-of-sight transmission is not easily accomplished, microwave will not be an attractive alternative and a wired (or fiber) network will be employed. Satellites, particularly small aperture satellite systems, are also increasingly popular where a company (or city department) wishes to have links to a national or international network. For example, both Phoenix and Houston have extensive microwave communications systems because the geography is flat. By contrast, the hills of Pittsburgh make microwave networks more difficult. Phoenix, it might be mentioned, recognizes its important geographical position for satellite communications between Europe and Japan. Los Angeles and Seattle consider themselves "gateways" to the Far East, not only in sea and air transportation, but in telecommunications as well. Houston and Miami promote themselves as offering links with Central and South American trade.

There is a general contrast among areas where large users avail themselves of local exchange company services because that company seems to be responsive, as contrasted with areas where large users go on their own.

There seems to be less bypass, for example, in Atlanta, where the local exchange carrier has a reputation as being aggressive. Also, where there is a pool of experienced telecommunications users, there tends to be a great deal of bypass (Houston is one example). One regional Bell executive reflected that where bypass is growing, the local exchange company has "not done its homework."

It was also noted that some examples of advanced telecommunications infrastructures came as a result of construction for a specific event, for example, the Democratic National Convention in Atlanta, or the 1984 Olympics in Los Angeles.

There has also been some assemblage of networks specifically for advancing links between R&D industries and universities as well as with government organizations concerned with development (discussed in Chapter 6). This example exists in New York as NYSERnet, in Pittsburgh as PRPnet, and CityNet in Houston which is a public service network for the city. In the R&D scene, many of these networks may tie in with such newly developed national networks for the promotion of scientific exchange and "technology transfer," as, for example, NSFNET (Williams and Brackenridge, 1990). Teleports, also an investment strategy, are discussed at a later point in this chapter.

2. Major Players in Telecommunications Growth

One would initially think that city managers and the local exchange company would be the main two players in the urban telecommunications scene. But this has diminished with divestiture and is changing even more as new players proliferate. First, much of the growing infrastructure is in the private networks of large users who themselves may command more "telecommunications power" in their own applications than does the city. These large users may have bypassed the local exchange company altogether. Second, there are interexchange companies vying for the business of large users and they may operate quite freely of any city influence. Third are other vendors of special networks for business, customer premises equipment, computer networks, PBXs, and the like. These companies, especially national or international ones, have a major influence in setting local standards. Finally, there are consumer groups who may fight for holding rates down, getting local dialing areas extended, or arguing against public network upgrade because it might favor only the large users. These people are vocal and have a definite effect.

Also, as mentioned in an earlier chapter, there is often a "balkanization" of telecommunications planning among the different agencies and branches of city government, each with its own priorities and standards. This is why it is not unusual that agency computers cannot "talk" with one another. Agency networks are often considered "turf," so outsiders are discouraged from interfering, all of which, of course, makes the overall city infrastructure inefficient and unwieldy. (In some cities it is next to impossible to describe all of the network services, as no central inventory is maintained.)

To counter this, some cities appoint telecommunications "czars" and may open a special office in the area. These range all the way from offices meant to coordinate cable television franchises to those that plan for the network needs of city agencies.

There are also "outside" regulatory players, the best known being at the federal and state level. But there are many examples where it is clear that federal or state positions on regulatory matters do not seem closely in touch with the urban market. Bypass is the major example.

A presumption of divestiture was that the local exchange market, especially the "local loop," would not be conducive to competition. Quite the contrary, because of the lucrative opportunities for doing business, especially with large users, there are all kinds of competition (VSAT, special fiber loops, cellular, and the possibility of personal communication networks, or "PCNs"). Many state utility commissions spend a great deal of time (and public funds) hearing arguments about competition in the local market. If addressed as a legal question, the debates (or hearings) could probably go on forever. As an economic or network study, empirical data can easily answer the question, and studies should be commissioned to do so.

Although one does not often find the title of "regulator" associated with a typical city administration, there are types of *de facto* regulation that take place, for example when assigning rights-of-way, enforcing cable franchise provisions, or monitoring quality of service. Added to this is the implication that when a city makes a large purchase or takes a particular position on a type of service to be provided, it is "setting standards." In brief, a city's regulatory influence is often a subtle one. We must recognize that many utility commissions hear more from residential advocates than large users because the former are the constituency of consumer groups and the latter are probably already off the public network. The consequence is a smaller rate base and, hence, fewer services or higher rates for urban residential customers and small businesses.

This is not to downgrade public interest groups typically concerned with residential services in urban communication; these groups serve an important balance in regulatory affairs. The problem is that they may not have a long-range view. If their constituents never have a chance to try new services, how will they know the possible benefits? We must be careful not to promote a stereotype that enhanced telecommunications options are only for the very rich. On the contrary, perhaps some should be part of the "basic service" available to every citizen. This is the topic of the California Intelligent Network Task Force, discussed in Chapter 12.

As we inquire into activities that relate telecommunications to economic development, there may be city development offices with particular interests in telecommunications, including quasi-governmental organizations such as a local chamber of commerce. All of the foregoing are often players at the urban levels of telecommunications planning and implementation, although the various positions of these players, their influence, and the likely consequences of telecommunications on economic development vary considerably from city to city.

Then there are cable television companies which may find their franchises threatened by the desires of local exchange telcos to enter into video and information services businesses. There is a reverse of this situation where the cable television provider may be offering data transmission services that encroach upon a local exchange company's monopoly. "Institutional nets," or special cable services for business or governmental telecommunications links, may directly represent a competitor on the local exchange level.

The interrelationship among these players often sets the "plot" for city activities in telecommunications competition, regulatory interpretation, and development. Some examples are discussed next.

Seattle was a particularly interesting example regarding the activities of the local Bell company, U.S. West. In this case, U.S. West was lobbying hard to prevent competition in its monopoly area, while attempting to identify a range of new services it might offer. Seattle is not only a rapidly growing economic area where there is an emphasis on aeronautics, computers, and high-tech manufacturing, but it also houses large corporations noted for their own innovations in telecommunications, if not operation of their own networks (e.g., Boeing).

Because of the number of players and the size of the prize, New York City's telecommunications environment is highly competitive. Businesses have increased the city's awareness of the need for telecommunications policies that encourage growth. Large users have been successful in changing harmful city policies through the New York City Partnerships. Metro-Tech is one example of all the players of the New York telecommunications game coming together—city, businesses, and providers—to encourage economic development in downtown Brooklyn.

Toronto was an example representing players still operating in a largely regulated environment. Locally, regulation seemed to be functioning adequately, with users satisfied with the services they received. But in the absence of a competitive environment, there was the growing question of whether industries doing business in less regulated environments (e.g., a U.S. city) were gaining access to less expensive telecommunications services. We found that some business services were more expensive in Toronto than in U.S. cities due to the cross-subsidies of business and citizen services in the former.

Although there were examples of cable television companies providing data services in Los Angeles, Minneapolis, and New York, it did not appear at the time of the studies that the cable player was a major threat to the local exchange provider. This competition could include the provision of institutional nets ("I-Nets") which were often included in the 1970s as a provision for gaining a cable franchise—that is, the serving of institutional customers with broadband services. By and large, we found that I-Nets were largely unexploited. Indeed, the opposite in terms of competition between the local exchange company and the cable television provider may be the case. Local exchange companies in Los Angeles, Houston, and Atlanta quite readily stated their ambitions for providing broadband telecomunications services to residential customers. Again, bypass appears to be an emerging point of contention between players in the urban telecommunications market. Many large users have taken their telecommunications systems "private," except for links into the local exchange networks. "Private" does not necessarily mean that they have built their own network but they may lease "virtually private" services from local exchange and interexchange providers.

It was also observed that telecommunications managers serving large users are often very much in communication with one another for information exchange, and mutual help, akin to computer "usage groups." Such groups were identified among large users such as Minneapolis, Boston, and Phoenix, as well as in Toronto. Large users typically seek control and flexibility in their networks as qualities sometimes more important than cost advantage. If the local exchange companies are to retain large users as customers, they will have to increase their flexibility and responsiveness, as well as options for "user control."

Finally, it is likely that we will see new players as telephone companies attempt to enter the information services business. Given the openings allowed by Judge Greene in this area, local exchange companies may provide transmission and gateway services, but as of this writing are still barred from providing the information itself. If the present regulatory constraints prevail, we may expect partnerships between local exchange companies and new players in the form of groups that have information to provide, such as newspapers, statistical information services, schedule and reservation services, shopping, or other videotex services. If the prohibition of Bell companies to provide the information themselves is lifted, we may see a closer integration of the information providers within the corporate structures of the Bell companies in the form of subsidiaries or partnerships. The local exchange providers have the network, and have long experience in customer billing systems, both of which make them worthy competitors in the business of home information services. As discussed in Chapter 7, such services may also include new players associated with on-line "yellow pages."

3. Cities as Telecommunications Planners and Users

Generally, one does not find telecommunications high on the list of city planning priorities. But in most cities, there are planning decisions directly affecting the infrastructure, as, for example, telecommunications rights-of-way. Another area is in contributing to the planning of the public network. Here, there may be a coordination between city plotting of areas of growth and encouragement of the local exchange company to install lines along with other utility construction.

Telephone companies have gained considerable expertise in predicting demand for public services, and these are usually put into play as a local exchange provider proposes expansion of the network to the regulatory authorities (because expansion costs will go into the rate base in most cases). Cities seemed least influential relative to the actions of large telecommunications users to develop their own networks. Large users, except for having to coordinate with the city on building code restrictions or certain types of right-of-way, generally seem to be proceeding with their private networks.

Some cities (Atlanta, Boston, Houston, Los Angeles, Miami) have moved quite aggressively in upgrading their internal administrative and emergency telecommunications systems. This includes not only more efficient voice services, but upgrading communications to high-speed data services, the offering of shared databases, and certain special services like video teleconferencing.

One challenge identified among the cities which were more aggressive in planning their own services was the inevitable politicization of the design and acquisition process. Not only do internal politics get involved in setting growth priorities but unfortunately also in selecting vendors, and sometimes especially so when the selection might involve a choice between the certificated local exchange company and an unregulated vendor. Managers often decried the problems of distributing requests for information and having to defend many details before city officials and councils in the planning process. The argument that advanced telecommunications systems will result in long-range cost savings is not always appealing to city councils with short-range political orientations.

On the innovative side are examples where cities are using telecommunications to solve a longtime problem. In Boston this was the traffic problem caused by frequently tearing up streets in order to upgrade buried utility lines, and without much coordination among the utility companies. Boston now has a sophisticated system for coordinating these activities, a "conduit policy" which reduces their problems. Although the new policy is primarily concerned with street management, it holds ramifications for the expansion of Boston's telecommunications infrastructure.

Minneapolis is one of the few cities to have launched a major task force initiaive in planning its urban telecommunications infrastructure. Its task force undertook to achieve three goals in 1985: (1) to inventory the city's telecommunications infrastructure, (2) to survey city businesses on their current and future telecommunications needs, and (3) to recommend policy changes to encourage the development of state-of-the-art telecommunications infrastructure. Since the largest employment sectors were in the services, trade, and high-technology manufacturing areas that depend upon advanced telecommunications, the city was sensitive to the needs of businesses and the threat that some would relocate. Although the study presented a set of rather ambitious generalizations, the consequence has been mainly for the city government to withdraw itself from trying to build public infrastructure, but, instead, to encourage coordination among others. The city government has also given considerable attention to its own internal telecommunications system.

The city of Los Angeles has been aggressive in attention given to its telecommunications infrastructure. As a planner, the Los Angeles City Department of Telecommunications regulates cable-TV franchises. As for planning in the emergency services area, Los Angeles has to deal with the constant threat of earthquakes. Consequentially, all of the emergency services in Los Angeles have been planned with extensive redundancy in their communications, such as microwave, separate radio systems, and alternative links through the public switched network.

As mentioned earlier, Los Angeles has also seen telecommunications of possible assistance in combating its air pollution problems. In the mid-1980s, the South Coast Air Quality Management District encouraged the city to require large companies (over 100 employees) to submit plans for meeting proposed average vehicle ridership for the upcoming years. This was backed by a threat of fines for those companies not complying. This led to considerable options for "telecommuting," or working at home. Although there is no evidence yet to support the expectations, officials estimate that telecommuting has the potential to reduce work-related automobile trips by 12 percent as of the year 2000. A pilot project has been organized under the auspices of the Central City Association Telecommunications Task Force. It is responsible for developing telecommuting pilots in downtown Los Angeles. The members of the subcommittee are carriers (AT&T, Pacific Bell, GTE), government agencies (Los Angeles City Department of Telecommunications, the California Department of General Services), and private companies (ARCO; Deloitte, Haskins, and Sells).

4. Telecommunications and Urban Economic Development

As with the state (Chapter 9) study, it was assumed that a major focal point would be to identify examples of telecommunications and economic development. For the most part, these examples represent initiatives rather than outcomes on the topic. There are numerous examples of where city planning in telecommunications may reflect goals related to economic development. This includes where a city might point to its telecommunications infrastructure as one of its many assets to lure a business to locate within its environment. As mentioned earlier in this volume, it is already well known that business relocation firms include an advanced (e.g., digital, fiber) telecommunications infrastructure on their list of desired urban qualities for information-intensive businesses (order taking, insurance, banking, transaction processing). Again, our ongoing generalization is that telecommunications is a "necessary but not sufficient condition" for development. It is not unusual for industrial recruiters in a city (from a development office, chamber of commerce, or task group) to include telecommunications vendors in their presentation of a city's assets.

Miami and Houston represented examples where development groups promoted telecommunications as a "gateway" to Latin America. And as already mentioned in this chapter, activities like Miami's LATCOM promote telecommunications businesses focused on Latin America. In order to attract telecommunications exporters to the city, promoters from private industry and civic organizations conducted efforts such as promotional trips to Telecom Geneva, the organization of the World Trade Center Telecommunications Committee, as well as producing publications promoting Miami services. Another development has included the development of a Greater Miami Chamber of Commerce Communications and Informatics Committee. These efforts also extend the public service areas relative to Latin America, as demonstrated by the Miami Children's Hospital production of Latin American satellite teleconferences.

The El Paso–Ciudad Juárez site represented an interesting contrary example to the use of telecommunications for economic development. There is very little initiative on the part of the city of El Paso to encourage special telecommunications needed to promote the maquila industries. Part of the reason for this is no doubt that the local exchange carrier (Southwestern Bell) is barred from the transborder telecommunications business. Another reason is that until recently, under a longtime agreement between AT&T and Telefonos de Mexico, their partnership was the only alternative for transborder communications. But now there are more options, including more international long-distance providers (e.g., US Sprint), as well as a variety of private microwave and satellite bypasses. Given a relaxing of Mexican regulations in this area in the Salinas administration (1989), we have seen private, satellite-based networks established which provide high-capacity communication of U.S.-based industry headquarters and suppliers with manufacturing sites in Mexico. Although maquilas are important to the economy of El Paso and Juárez, and telecommunications is an important infrastructure component, neither city, itself, has done much in this area.

Ciudad Juárez provided an even more striking contrast between infrastructure as used to benefit a population as against that for large users. Telephone penetration of Ciudad Juárez at the time of this study was approximately nine telephones per 100 inhabitants (the U.S. average was 80 per 100). Further, the price or residential telephone service is beyond the means of the average Ciudad Juárez household. We saw little use of the public telecommunications infrastructure in this Mexican city for large users of its everyday citizens.

Seaport examples included Seattle and Los Angeles offering communications and data processing services to port customers. The port of Seattle offers services that facilitate communication of inventory, shipment, and customs data. Los Angeles, through its World Trade Center Association, has a sophisticated network service that matches sellers and buyers in the global economy. The nearby port of Long Beach also participated in that project.

NOTES ON TELEPORTS

The Concept

"Teleports," drawing on thoughts about seaports or airports, would seem an important investment for a city. But they turned out not to be as important as originally thought in the study. As of this writing, teleports have probably received more publicity than they deserve, some of them being nothing other than a concentration of antennas. Reports of teleports in Minneapolis, Phoenix, and Boston have shown them not to be of marked importance to local infrastructure, at least at this point. The New York/New Jersey Teleport as well as the Houston Teleport were useful facilities for promotion, and some users did cite their advantage.

New York/New Jersey Teleport

The New York/New Jersey Teleport is located on Staten Island and is a joint project of the Port Authority of New Jersey and New York. After becoming operational in 1985, it achieved an early success with the completion of a 150-mile fiber optic network connecting the Teleport to much of New York City and parts of New Jersey. An attempt has also been made to revitalize the economically depressed areas of Queens and Brooklyn bringing them into the network via fiber optic links. With this network available to them, corporations could move their back-office functions to these areas, where real estate and other costs are lower. If the project succeeds, New York will have attained three benefits: a state-of-the-art infrastructure for the city; a retention through relocation of the firms now fleeing the city; and a revitalization of the city's economically depressed areas.

The telecommunications component of the Teleport is managed by Teleport Communications, a partnership between Merrill Lynch and Western Union. The real estate component is managed by the Port Authority of New Jersey and New York. The basic product offered by Teleport Communications is "antenna slips," which can be described as "a high quality, high security environment with support facilities (e.g., emergency power) for satellite earth stations and ancillary equipment." Such an environment enables the customer to overcome the microwave congestion and new connection difficulties in New York City. A company may choose to install its own equipment in the Teleport, share it with others, or use Teleport Communications' equipment—thus providing economies of scale.

The New York/New Jersey Teleport project was designed with the main objective of stimulating economic development in the New York metropolitan area by strengthening the telecommunications infrastructure of the region. The Port Authority is in the process of completing an in-house study measuring the economic impact of the Teleport. Expectations remain high. In New York, a main component of Teleport is that it includes an office park and it will attract, to a relatively undeveloped area, a million square feet development initially and as much as two million square feet eventually. In turn, this office park is expected to house about 3,400 jobs and create another 2,000 jobs in local employment. This could result in an annual payroll in excess of $100 million on Staten Island. Revenues to the localities could total about $10 million a year. All of this is extremely welcome as economic develpment.[2]

Bay City Teleport

The Bay Area Teleport (BAT), in the Alameda area of California, is a private investment project in this competitive, dynamic region. The Teleport, developed by Doric Development and Northern Telecom, is linked to the largest antenna facility in the world, the Niles Canyon Earth Station, and a 3,600-mile regional digital microwave network. This network, linking major cities from Sacramento to San Francisco to Silicon Valley, allows users to bypass the local exchange companies and connect through the Teleport, long-distance companies, and the antenna earth station directly to Europe and Asia.

The Teleport became operational on February 2, 1986. Although the real estate development to which the Teleport is related, the Bay Area Business Park, has been under construction since the 1970s, the teleport itself was born in the days just prior to and immediately following the divestiture of AT&T, an era which encouraged the Teleport's founders and others to take advantage of new investment opportunities in telecommunications.

The Teleport's claims to economic development hold that businesses will be able to communicate with voice, video, and data to both Europe and Japan and all points in between on the same day and through the same location. Furthermore, some observers believe teleports to be future sites of company relocations.

The San Francisco Bay Area, home to 34 of the Fortune 1000 businesses, spends $200 million annually on telecommunications. In addition, Silicon Valley, a high-technology hub, is just south of San Francisco. Teleport personnel believe that a $50 million market is available to them in the next few years. Customers include the major domestic providers of long-distance telecommunications, financial institutions, government offices, and other businesses and international telecommunications firms like US Sprint International, ITT World Com, and COMSAT International. While there is no evidence to suggest that major companies locate next to the Teleport just to take advantage of the telecommunications technology, most of the firms in the Teleport's business park and real estate complex subscribe to their services. In an indirect way, then, the Teleport has already had an impact on telecommunications use in northern California.

Bypass is not the only reason for the Teleport's existence. Phase I of the project is the business park, which includes the Teleport and companies like Litton Industries Integrated, DEC, and RAND Information Systems. Phase II includes a world trade center, a bank, and incubator space for the startup of high-tech industries. Phase III is an international research complex that will include a new product demonstration center, the University of California Research and Patent Center, and a branch of the Lawrence Livermore National Laboratory.

The main projects of these phases are expected to be completed within the next ten years. Whether the Teleport is merely taking advantage of existing business or whether it is generating new economic development is difficult to say, but most agree that it has been influential. The three components of the teleport—the intelligent park (local area network, shared tenant services), the intelligent region (microwave towers throughout northern California from Sacramento to Silicon Valley), and the earth station antenna complex—are seen by the Teleport as an emerging global center for telecommunications and economic development.[3]

OVERALL RECOMMENDATIONS ON CITIES

Again, the findings of the urban study tended to fall into four main areas. In a broad view, generalizations in these areas were as follows, and the policy and planning recommendations derived from these are summarized in Table 10.1.

First, there is abundant evidence that urban stakeholders are increasingly recognizing telecommunications as an important infrastructure component. The idea of making strategic investment in this component makes sense to most planners.

Second, divestiture and the bringing of competition to the telecommunications marketplace has greatly increased the number of "players." Planners may have to deal with many different vendors and perhaps an overall one to coordinate the smaller ones. Despite the assumption of the local exchange company as the natural monopoly, competition seems to be growing rapidly, especially for business and public service telecommunications applications.

Third, cities very widely in their skills or interests in telecommunications planning. But there are visible areas of increased interest, not only in planning better telecommunications for the "city as a user," but in planning infrastructure investments to attract or retain businesses.

Fourth, although the exact relation between telecommunications and

Table *10.1.* Summary Recommendations

(Adapted and expanded from Schmandt et al., 1990).

Urban telecommunications planning should be approached as a *strategic investment* and not just maintainence of another city utility. If cities are to compete in development and in the delivery of services, they have to become more aware of advanced telecommunications as an option for investment and planning. Cities seeking growth should explore how telecommunications is often seen *as a necessary but not sufficient condition* for economic development.

There needs to be a better understanding of the implications of the coming *intelligent network for urban applications.* Many city influentials are unaware of the applications of modern telecommunications. Stakeholder groups should be organized and given seminars by the major telecommunications providers.

Urban administrators can act as facilitators and coordinators for telecommunications development. The city does not have to build the network, although this is an option to consider. The above-mentioned seminars can extend to cooperative planning sessions. Such groups should stress consensus-building among stakeholders. Urban representatives should try to *increase their presence on state and federal policy-making bodies.* Much of the concentration of telecommunications investment is in cities, yet regulation is located on state and federal levels.

Cities should consider how they can add to their own expertise in urban telecommunications. A *Telecommunications Office* can integrate planning on the cities telecommunications uses, cable franchises, issues of right-of-way and taxation, as well as use of telecommunications expertise to attract new businesses or revitalize old ones. Modern cities need to add constantly to their expertise in telecommunications technology.

City governments should be able to describe and stay abreast of the overall *telecommunications capacity of their infrastructure.* To this end, consideration should be given to establishing a "clearinghouse" of information on the city's infrastructure. Awareness of a city's telecommunications capacity will facilitate economic development programs. (It was noted that such special features as cable-TV I-Nets are often underused.)

City planners should *recognize the de facto policy-making role* inherent in decisions they make about telecommunications (e.g., in right-of-way, franchising, taxation, large purchases). Although most city governments are not directly involved in drafting and regulating telecommunications policy, their decisions and their behavior can be interpreted as guidelines to, and tests of, policy.

Although the research team endorses the position of the National League of Cities, that telephone companies be allowed to *compete in the urban cable market*, cities should take great care not to suffer losses in service and opportunities as these players vie for the consumer market.City planners need to be especially aware of the changing regulatory position on cable franchises.

Table *10.1.* Summary Recommendations (*con't*)

City telecommunications planners should know that the traditional definition of *universal service* as being "dial tone" may be *giving way to citizen access to advanced services*, including videotex. Therefore it is necessary to review traditional premises in telecommunications planning and policy making.

Cities should be aware that advanced telecommunications enables them to be *increasingly competitive in the growing global marketplace*. Thus, what may have appeared to be local telecommunications issues may become international ones. Cities, through telecommunications, may become more independent of local and regional economies.

Cities should encourage *data gathering on important telecommunications investments* and their consequences. To date, such data are nearly nonexistent. We will never know much more about the relation of telecommunications with urban development unless some stakeholders take the initiative to commission studies.

Although *teleports* may seem an attractive investment, it is important that developers *examine existing examples*. The myth has exceeded reality in some cases.

Training programs in city government should include telecommunications planning and management. Cooperation should be sought with nearby colleges and universities to teach telecommunications planning and management. Cities, their different agencies, and other large users should consider methods by which different users can share in the same training sessions. Similar programs could be extended to other area users, employment agencies, displaced workers, and schools.

economic development remains a difficult study for concrete and quantitative data gathering, example after example of initiatives provides anecdotal evidence of its importance. At least we know that the absence of suitable telecommunications can be a negative factor for development (if not a disaster). Finally, teleports are not necessarily a key investment; a city should examine examples before making plans in this area.

■   ■   ■

Any city serious about its economic development will have to take an active role in examining its telecommunications infrastructure, determining its own policy and planning priorities, and putting in place some mechanism (official, office, advisory group) to carry out these activities. Cities, too, should look to partnerships with their local exchange companies. Too much bypassing could be to the detriment of both.

11

•

Rural: Developing the Land Between the Cities

"Go ahead, call somebody," says Jimmy White, wheeling his blue sedan down a farm-to-market road on our way to visit a farm near Dalhart, Texas. This is wide-open range country about 80 miles north of Amarillo, where the land is nearly flat from horizon to horizon. You'd think you were doing a scene in the movie *Giant*, except Jimmy is driving a 1990 car and he holds the latest model "Flip Top" phone for me to try. Jimmy is excited about the rural uses of cellular phones, which he ought to be because the XIT Telephone Cooperative that he heads owns most of the franchise. (You see "XIT" everywhere here; it's the cattle brand from what used to be the largest ranch in North Texas.) Jimmy describes some of his best customers and how important telecommunications is to their rural businesses ("the feedlot has a phone in each of its trucks") and residential life ("nobody has to feel alone anymore, no matter how far out they live"). "This is changing our way of life out here," explains Jimmy. "It used to be that most of our phone traffic was just after dark when folks came in from the fields. Now its like in town; it peaks around two in the afternoon." As he starts to dial his office, Jimmy explains that a farmer or cattleman can call whenever and from wherever he wants—simple as that. But the phone rings first. It's Larry Kemp calling from his tractor; he'll be about 20 minutes late for our meeting because he's chasing down a stray calf.[1]

Can telecommunications help to revitalize the economy of rural America? Can we use it in innovative ways to attract new population away from crowded and overbuilt cities? Are there new options for developing "the land between the cities?" In this chapter, we review some of the new ideas for using telecommunications as an investment in the rural economy and lifestyle.

■ ■ ■

TELECOMMUNICATIONS AND ECONOMIC DEVELOPMENT IN RURAL AMERICA

"Nobody Lives in the Sticks Anymore"

Contrary to some city dwellers' image, life in modern rural America is more than farms, fences, cows, and broken-down pickup trucks. This stereotype is no more valid than the supposed country person's image that life in the city is one of only traffic jams, pollution, and protecting your hubcaps. Most Americans who live outside of metropolitan areas (defined by the U.S. Census as population concentrations of 50,000 or more) work in manufacturing jobs and not on the farm. Even a majority of "non-metro" Americans who do farm work also have another job. Television has merged urban and rural views of popular culture, a point that led TV researcher George Gerbner to reflect in an interview on the "Today Show" that "nobody lives in the sticks anymore."

According to studies by the U.S. Department of Agriculture,[2] we can generalize the following about rural America in the information age:

- Of the nation's 3,097 counties, 2,383 are "nonmetro" according to the above definition.
- When characterized by their income patterns, the nonmetro counties reflect economic diversity among manufacturing (18 percent of the counties), retirement (17), farming (16), government (12), federal lands (8), poverty (8), and mining (4), with the remaining 17 percent unclassified.
- In 1985, the rural poverty rate was 18.3 percent as compared with 12.7 percent for the metro population; while the metro poverty rate fell in the early 1980s, the nonmetro rate did not.
- High school completion in rural area has remained about 10 percentage points behind metro statistics since 1960; the gap for college completion has also remained relatively consistent.

Despite the lag of education and business in rural America, many small-city and village dwellers prize their lifestyles. In much of rural America a young couple can still buy a house, enjoy crime-free schools, and go hiking in five minute's time in a countryside with clean air and water. There are many that prefer a "country" lifestyle—to stay where they are, or to move from an overcrowded and overtaxed city to a more easygoing way of life. The barrier is, first of all, the rural economy. Even if the cost of life is less, this means nothing if you cannot generate a basic income. Second is the problem of infrastructure. A town without a hospital or a village without a doctor is not attractive either to retirees or couples raising a family. Well-educated families are hesitant to enroll their children in schools with limited curricula and less experience in preparing a college-bound population. Rural security can be a problem when you consider that there might not be a fire department or sheriff within a reasonable distance in the case of a potentially disastrous emergency.

As posed at the outset of this chapter, the question is whether telecommunications can contribute in any way to solutions for these problems.

Revitalizing Rural America

Before considering telecommunications applications, it is important to understand the changing views on which economic and infrastructure investments may be of maximum benefit for development. For most of modern times, the development of rural America was seen as agricultural development. The economic policy was a farm policy, and mostly in the hands of the U.S. Department of Agriculture. But times have changed. As drawn from U.S. Department of Agriculture research and policy statements (1987), there are fresh approaches to considering the future of rural America; in summary form, these include:

Separation from agricultural policy. Rural development policies must become uncoupled from agricultural policy. The long-term decline in agricultural employment, due to changes in rural economic emphases, loss of competitiveness, and worker displacement by technology, calls for a separate policy for overall rural economic development. What are developmental policies and objectives separate from agriculture—as in manufacturing, recreation, retirement communities, or new service industries?

Protection. Being shielded by trade barriers from a competitive global economy is in many ways ultimately self-defeating for rural areas. It may stifle business creativity and aggressiveness, as well as other bases for developing or sustaining competitiveness.

Subsidies. Subsidizing existing businesses in rural areas is not a promising policy because many of the economic problems of farming, mining, and rural manufacturing result from existing competitive disadvantages nationally and, increasingly now, globally. The need is to identify potentially stronger business opportunities that can benefit from rural locations. The bottom line is what can be done competitively.

Education. This is a critical investment for enhancing human potential in rural areas. Many rural businesses now require a more highly trained work force. But if the work force is upgraded, it is also important to have jobs available lest the population leave the area in search of better opportunities.

Movement toward local and state initiatives. Many experts agree that initiatives for development will be most likely and have the greatest chance for success if mounted on local, state, or regional levels. This is a mandate for local leadership.

Infrastructure. Transportation, water, power, or waste disposal have often been traditional problems of infrastructure that inhibit rural economic growth. In the information age, telecommuncations is growing as an important new element of infrastructure.

Telecommunications. This and the ability to use information technologies are requisites for many directions of business growth. Investments in the uses of telecommunications can also benefit rural education.

In all of the above, we may be considering more than small-town or city maintenance or revitalization. As cities become overcrowded and inefficient, we could well turn to new goals in rural development, namely the creation of new residential and business areas in what we have called "the land between the cities." Where do telecommunications investment fit within this picture?

Why Telecommunications?

In many ways, telecommunications addresses the two interrelated aspects of the "rural disadvantage," namely: distance and low population density. For some services and needs, telecommunications can make distance irrelevant; and by overcoming distance, telecommunications can create new economies of scale. As in several of the small business examples (Chapter 5), a company can coordinate its activities with suppliers and customers through telecommunications; this, combined with "parcel" shipping services, make rural locations feasible if not attractive. By reaching out regionally or nationally via telecommunications, a sufficient number of customers can be reached to form a viable market. Distance learning is a parallel public service example. Courses can be initiated from anywhere, so long as the school is on the network. If a number of spatially separated schools can join in using a "course," then there is a sufficient economy of scale to offer it.

In overview, some of the major benefits accruing from overcoming distance and creating economies of scale for rural areas, and likely applications for the intelligent network, include:

- creating large markets for small business
- location and coordination of branch plants
- operation of telemarketing
- enhanced business (including agricultural) information services
- better market information for farmers and ranchers
- more opportunity for small-town revitalization
- more chances to promote and coordinate tourism
- improved education through distance learning
- coordination of rural health care delivery and administration
- improved coordination of law enforcement
- increased ability of small towns to cooperate with one another
- better emergency services
- automatic monitoring of natural resources (e.g., water table)
- improved scheduling of transportation
- more alternatives for home information and entertainment
- less isolation for remote rural residents
- improved opportunities for village administration

Need for Developmental Studies

We have already discussed in Chapter 3 how third world studies (e.g., Hudson, 1984; Saunders, Warford, and Wellenius, 1983) of telecommunications and rural development carried implications for industrialized countries. Heather Hudson (1987) addressed this issue in a report on applications in North America, although most of these were more reflective of remote and undeveloped rural areas than of the bulk of rural America. Rural sociologist Donald Dillman (1985) at Washington State University is a major observer and researcher on the social impacts of information technologies in rural America. He has written on the growth of five particular features of the emerging rural information structure:

- greatly expanded telecommunications
- equipment needed for the productive use of these capabilities

- information technologies that come embedded in tools and materials
- a rapid physical delivery system
- the capability of individuals to use the new information technologies.

In rural environments, the challenge is not so much the feasibility of installing new information technologies as it is the fact that the details of adoption in business, educational, and cultural sectors are so little known. Dillman acknowledges that the rural sector lags behind in adoption of these technologies, not only because of economic barriers but also because rural educational levels are lower than those found in urban area. By overcoming the isolation of rural areas from information sources, message and trans-action systems, and educational resources, telecommunications has the potential to effect positive change. These resources and capabilities can open new advantages for rural businesses whose interests reflect a wide variety of endeavors in addition to agriculture.

As mentioned above, rural education may benefit from the use of telecommunications to deliver new learning materials to remote schools. And the social experiences of rural dwellers—already profoundly affected by television—are again being newly influenced by the increasing use of receive-only satellite dishes.

ASPEN INSTITUTE AND FORD FOUNDATION INITIATIVES

In July 1988, some two dozen rural development specialists, telecom-munications researchers, and representatives of the Aspen Institute's Rural Poverty Program and the Ford Foundation met for three days to seek insights into how new investments in telecommunications might benefit the rural economy and population in various ways (if it would at all). As summarized by David Bollier,[3] the group concluded that telecommunica-tions, although certainly no panacea for solving rural problems, could be better employed. Among the action suggestions, particularly as advanced by David Dillman, were three levels of upgrade for rural telephone service. The first was to extend service to those who still did not have access to the network either because of extreme remoteness, poverty, or both. A second was to upgrade services to small businesses, including data services (as discussed in Chapter 5). The third was to use telecommunications to improve the delivery of education to rural areas.

Another main consequence is that the conference set several agendas for compiling more detailed information on the potential for rural telecom-munications investment and to frame specific policy issues. Most visible of these was the publication in the following year of *Rural America in the*

Table *11.1.* Aspen Group Recommendations

(Adapted from Parker et al., 1989, esp. p. xii.)

1. Make voice telephone service available to everyone.
2. Make single-party access to the public switched telephone network available to everyone.
3. Improve the quality of telephone service sufficiently to allow rapid and reliable transmission of facsimile documents and data.
4. Provide rural telephone users with equal access to competitive long-distance carriers.
5. Provide rural telephone users with local access to value-added data networks.
6. Provide 911 emergency service with automatic number identification in rural areas.
7. Expand mobile (cellular) telephone service.
8. Make available touch tone and custom calling services, including such services as three-way calling, call forwarding, and call waiting.
9. Make voice messaging service available via local phone calls.
10. Help rural telephone carriers to provide the telecommunications and information services that become generally available in urban areas.

Information Age: Telecommunications Policy for Rural Development, by Edwin B. Parker, Heather E. Hudson, Don A. Dillman, and Andrew D. Roscoe, all participants in the original conference. The volume was prepared under the same sponsorship as the original conference, namely the Ford Foundation and the Rural Economic Policy Program of the Aspen Institute; it has been widely distributed to rural policy makers.

The volume provides a summary of status quo conditions on the layout of rural public telecommunications, then goes on to advocate specific upgradings of telephone services, as summarized in Table 11.1. Finally, the authors propose several action strategies for gaining these upgrades, the main one being to increase the availability of Rural Electrification Administration (REA) low-interest loans to pay for the improvements.

A STUDY OF RURAL BUSINESS
AND PUBLIC SERVICE APPLICATIONS

Background

This was the third in the series of Policy Research Projects reported in the state (Chapter 9) and city (Chapter 10) chapters. The rural project (Schmandt, Williams, Wilson, and Strover, 1991) was prompted by two factors, the recognition that rural areas of the country are suffering economic duress, and the widely held expectation that telecommunications offered opportunities for ameliorating some of the difficulties facing sparsely populated and remote parts of the country. The project sought answers to three primary research questions:

- What applications of innovative telecommunications occur in rural areas?
- What role does telecommunications play in rural innovation and economic development?
- What are the policy and development implications of innovative telecommunications-based applications?

To address these questions, a comparative case study approach was adopted. This involved research into, and visits to, 37 sites reflecting public and private institutions that either deliver social services, run businesses, provide telecommunications services, or promote economic development in rural areas. The sites reflect the economic and geographical diversity characterizing rural America. Their economic profile includes communities whose economies are dependent on agriculture, manufacturing, retail, tourism, and public services. The work was done as a year-long project during 1989–90. Sites were divided into four categories for reporting and interpretation: (1) businesses, (2) public service applications, (3) small rural telcos (plus several alliances and a power company providing telecommunications), and (4) rural communities using telecommunications as part of their development strategy.

In the largest view, the cases examined in this study suggest at least three components to rural innovations:

- Many of the innovations depended on discovering ways to aggregate demand, thereby creating economies of scale typically absent in rural areas.
- Creating partnerships provided an institutional mechanism to exploit potential scale economies and to gain a critical mass of resources to do so.
- Leadership, sometimes local and sometimes regional, was critical in most cases; specific individuals can be credited with having the vision and the talent to formulate and execute novel plans.

Although the detailed recommendations from this study can be found in the original report (Schmandt et al., 1991), many are included in those given later for this entire chapter. Highlights of the findings are next discussed in terms of the four site categories.

Table *11.2*. Summary of Rural Policy Recommendations

(Adapted and expanded primarily from the author's earlier policy paper [1988b], the original Aspen conference [Bollier, 1988], the Rural Policy Research Project [Schmandt et al., 1991], and the ongoing study of four small cities [Strover and Williams, 1990].)

Given an understanding of the distinctive needs of rural populations, we should *create new incentives in our telecommunications planning* and regulatory structure to facilitate rural development. This means promoting access to intelligent network services.

We need to understand more about the *complementarity of rural economic development and the use of telecommunications*, including quantitative assessments.

We need first to evaluate current infrastructure and *set priorities for upgrades* (e.g., single-party service, etc.), then plan for development of the intelligent network.

We should determine *new strategies for financing rural telecommunications* including consideration of an expanded REA program for overall development of a rural area's network infrastructure.

We need to review critically and *revise existing U.S. telecommunications policy* that either overlooks, or is at odds with, rural interests. This means better liaison between state PUCs and development agencies.

Service providers should be encouraged to *develop and disseminate company policies on rural services* and their intended role in economic development, including assisting in public service planning.

1. Doing Business in Rural America

Sites in this category included the following organizations: Weyerhaeuser (Washington), John Deere (Illinois), Wal-Mart (Arkansas), Cabela's (Nebraska), and EMERG (Nebraska).[4] In these cases we sought to study how the extension of advanced telecommunications to rural areas enabled businesses to locate in such areas, or enabled other traditionally rural industries to be more efficient by competitively exploiting certain rural resources and markets. Although many details can be found in the study report (Schmandt et al, 1990), the following questions and briefly summarized answers provide an overview of findings.

1. What advantage does telecommunications provide for businesses operating in rural areas? Weyerhaeuser, John Deere, and Wal-Mart could not operate on the scale they do without the use of advanced network services. They were an excellent example of how telecommunications applications in large businesses can bring modern, urban management methods to rural areas. (A case study of Wal-Mart was included in Chapter

4). Small businesses like Cabela's and EMERG are examples of businesses built upon telecommunications capabilities. The vast market available to Cabela's is facilitated by catalog mailings and in-bound, toll-free numbers for customer ordering. EMERG in some respects is an example of how an "administrative subcontractor" can serve clients via network services. All provide examples of how telecommunications, by overcoming distance and allowing economies of scale, can facilitate rural businesses.

2. What are the effects of management philosophies on choosing tele-communications systems? It is noteworthy to observe that the large users are deploying their own networks in rural areas. Wal-Mart and John Deere use VSAT, and Weyerhaeuser is a mix of public and private lines. This suggests, in part, that if public telecommunications providers do not have the services or cannot compete with private alternatives, we will see most large rural users off of the public network, and their revenues lost to the public rate base. The challenge is to develop public alternatives that can compete for large customers and thereby create income that will support bringing advanced services to the countryside. The Cabela's and EMERG case studies also revealed problems in gaining access to needed services.

3. How have these industries affected local economies in rural areas? All examples were thought (by local residents, officials, other businesses) to have benefited their local economies. This would hold even for Wal-Mart which, while making it difficult for "mom-and-pop" businesses to com-pete, does bring an overall greater aggregate business to its locations than the small stores did collectively before.

2. Public Service Delivery

Sites in the public service area included Geisinger Medical Center (Pennsylvania), Minnesota Distance Learning Network (Minnesota), TI-IN (a distance learning service from Texas), and the U.S. Department of Agriculture (Washington, D.C., Florida, Iowa, and Texas). These cases illustrate how telecommunications resources are being used to overcome distance barriers for management as well as for medical, educational, and information service delivery. The following questions and brief summary answers provide a sample of this research area:

1. How have rural communities used telecommunications to gain access to public services? Networks bring new economies of scale in the delivery of human services. The Geisinger Medical Center is essentially a cluster of networks that greatly facilitate coordination among widely dispersed clinics, individual physicians, testing laboratories, research services, and

financial and insurance services (more in Chapter 8). TI-IN or the Minnesota Distance Learning Network allow instruction to be disseminated over wide areas, overcoming the problems of rural distance. The USDA is already well known for its use of network services, not only to coordinate its hundreds of field offices, but as an "information conduit" to the county agent, and eventually farmer or rancher.

2. Which public and private institutional arrangements are most conducive to rural public service delivery? The research on this question was essentially a comparison between the delivery systems of TI-IN, a national (one-way audio/video, and audio return only) satellite-delivered instructional system, and the Minnesota Distance Learning Network, a regional switched network (two-way audio/video). As discussed in more detail in Chapter 8, both systems have their advantages. TI-IN can inexpensively cover anywhere a satellite dish can be located, whereas the Minnesota system requires expensive networking. Yet the latter gives the advantage of two-way (like video-conferencing) instruction, which teachers feel comfortable with as a classroom extension. As concluded in Chapter 8, the best arrangements will probably be effficient combinations of both services. It was noteworthy that both services were initiated in their early stages without state assistance.

3. Small Rural Telecom Providers

An important consideration in U.S. rural telecommunications is that many areas are served by the nation's some 1,200 small independent telephone companies and cooperatives. Mostly these are areas that the original Bell system did not "wire" because population was so sparse, so in many cases ranchers or groups of farmers either started a company or a cooperative. Subsequently some of these areas grew in population and the "local telco" was bought either by the Bell company or large independents like GTE, United, or others.

These remaining small companies—like Jimmy White's XIT in the opening anecdote—are critical for the success of a rural telecomunications policy, not only because they often represent very small or remote areas but because they often aggressively promote local development. Most are owned by individuals or families who reside in the area, and, of course, a "co-op" is owned by its subscribers.

Too often these companies are either overlooked in policy studies or they are erroneously lumped in one supposedly homogeneous group. As with most human organizations, the smaller they are, the more they reflect the attitudes and abilities of individual developers, owners, or "champions" as one hears from some communities. Although Bell companies and large independents serve more rural customers than these small organizations, the project focused on the latter because we know relatively less about them. On the positive side, however, is that when one listens to anecdotes about successful examples of rural development using telecommunications, these companies are often involved.

In our project we studied selected small companies and also some cases of "alliances" for special network or service development. We also included one power company as an example of the ambitions of some in this business to provide telecommunications services. Our case studies included cooperatives (Eastern New Mexico, New Mexico; Eastex, Texas; Mid-Rivers, Montana; XIT, Texas); small telephone companies (Big Bend, Texas; Bretton Woods, New Hampshire; Clear Lake, Iowa; Kerrville, Texas; North Pittsburgh, Pennsylvania; Taconic, New York); regional networks (Iowa Network Services, Iowa; PalmettoNet, South Carolina); and an electric cooperative (Cotton Rural, Nebraska). Again, the results are described in terms of the main questions and very brief summary answers. More details can be found in Schmandt et al. (1991).

1. How do small rural telcos interact with their communities? For the most part, they interact in a very responsive and "closely knit" fashion because their owners (or members in the case of cooperatives) are a part of the community. On the positive side are that these service providers understand the needs of their community, have small-town personal relationships with their customers, and in many cases have a relatively upgraded network because they have been able to obtain low-interest REA funds. It is not unusual to see a small-town telco owner or manager as one of the local leaders serving in the chamber of commerce or other developmental groups.

On the negative side is that, due to their small size, it is very difficult for them to know the many modern applications of telecommunications in business and public service. Thus a town with a digital switch might not benefit from all of its services because local businesses do not know about new applications, and the telco has no staff to suggest them. And, of course, there are small telcos that are essentially "holding actions." They see themselves as a traditional utility with the mission of providing services when requested, but otherwise do not operate in a developmental mode. Groups like the Organization for the Protection and Advancement of Small Telephone Companies (OPASTCO) try to counter this problem.

2. What challenges are faced by rural telecommunications companies? If you talk in depth with small telcos or go to their association meetings or

workshops, the greatest fear is that they will loose access to outside funds that support their high-cost operations. (This is the form of high-cost pools paid into from urban operations, intrastate long distance, and the like.) As we move from a regulatory paradigm to a competitive one, there are many small telcos that could never "drive prices to costs," because the cost of local service would outstrip the ability of local customers to pay. Policywise this is a threat to traditional universal service.

What do they do about this? Some small companies and cooperatives are already developing lines of business outside of being a local exchange operator (partnering in cellular franchises, going into cable TV if under 2,500 lines, or selling equipment). If they are sufficiently diversified, the new revenues might allow them to replace loss of outside funds that support high costs. Another helpful trend is that modern switches and transmission technologies eventually lower the costs of operation and maintenance. A digital switch allows subscribers to be charged at the central office and provides new methods of troubleshooting, thus reducing the expensive time required to send service personnel out into the field ("windshield time" in the trade).

Nevertheless, any new rural telecommunications policy development has the high-cost problem to contend with, and the small telcos will be at the heart of the issue.

3. How do rural telcos approach innovative services? Admittedly, the small telcos selected for the present study were already known to be involved in some type of innovation or local development initiative, so the results are not a national "sampling" of the degree of innovation evolving in the industry. On the other hand, there were many examples in the present study that illustrate methods of innovation, and these will likely spread to interested parties in the industry. Among the innovations are digital switches, rural cellular services, cable TV, alliances for providing equal access to long-distance providers, distance learning networks, and deployment of fiber.

Three important factors assist the development and diffusion of innovation. One is that REA funds can provide low-cost financing for same. A second is that many small telco owners or managers frequently attend state and national conferences where ideas and strategies for innovation are shared. And the third is that the more successful small companies often have a "champion" in charge, a person who is professionally on the lookout for new opportunities. A final note is that rural power companies have ambitions in the telecommunications business, and it will be important to track their progress and potential contribution to development.

4. Telecommunications and Community Development

In rural studies, it is difficult to grasp the relative role of telecommunications without seeing applications in a context relative to the needs, problems, economy, and attitudes of a community or region. There may be no innovations in telecommunications because there are no opportunities for its application; the local attitude may be negative to growth or change. Or a rural area may be dominated by a diminished mining or oil economy, or a growing manufacturing one. In the present research, we located communities where there were potentially interesting generalizations to be made—positively or negatively—about the use of telecommunications for development. A summary of this phase of the project can best be given by a few notes on each community.

Dahlonega, Georgia. Located 70 miles outside of Atlanta, this community has made a point of its early installation of a digital switch and fiber services as an attraction for new businesses. It also uses telecommunications to promote its "adjacency" developmental relation with Atlanta.

Eagle Pass, Texas. The economy of this mid Rio Grande border city of about 28,000 is highly tied to its large Mexican neighbor, Piedras Negras (around 250,000). Yet telecommunications links in this important relationship are abysmal. Because it is barred from interLATA markets, the local service provider, Southwestern Bell, can do little to help in transborder telecommunications. The most highly creative (and often illegal) forms of bypass result. Policy for telecommunications in the border region will have to change drastically if the United States and Mexico are to institute a successful free trade agreement.

Hailey, Idaho. This state had earlier encouraged upgrade to digital switches and fiber trunks to promote new types of businesses and services. Hailey houses a consulting engineering company (Power Engineers) whose personnel were attracted to the outdoor lifestyle (Hailey is 15 minutes from the famous Sun Valley ski resort) and a telecommunications infrastructure that could support their widespread communication needs.

Kearney, Nebraska. This community has received national recognition for its promotion of telemarketing as one line of business development. Its success has depended upon the ability of local entrepreneurs to gain access to advanced network services which, once in place, attract other businesses. Also, the local state college offers cooperative relationships in the form of part-time workers and a telecommunications management degree program.

Ottumwa, Iowa. This community of about 25,000 has had a successful strategic plan for economic development, which includes telecommunications services and distance education.

The foregoing community studies also influenced two new projects focusing on communities, the reports of which will be available at a later date.[5]

SPATIAL PATTERNS OF
TELECOMMUNICATIONS AND RURAL DEVELOPMENT

Research across the foregoing studies and others has prompted several generalizations about the spatial patterns relevant to applications of telecommunications in rural development. On a broader level than a specific business or social service is how rural communities may attain economies of scale by linking their development with other communities; in most of these cases, telecommunications can play an enabling role. As illustrated in Figure 11.1, we have studied examples of newly developing "hubs," "adjacencies," and "alliances." We can summarize examples of these patterns briefly as follows:

Hubs. Sometimes a town or small city can become a "hub" for services to surrounding rural areas. There are sufficient economies of scale so that the hub may also serve surrounding rural residents with jobs, medical services, advanced education, financial services, and major shopping. Examples we have studied include Kearney, Nebraska, which among other characteristics has developed telemarketing jobs that draw employees from the rural population. Another is Demopolis, Alabama, which may be on the verge of becoming a hub if it can further develop

Figure *11.1.* Spatial Patterns of Rural Development

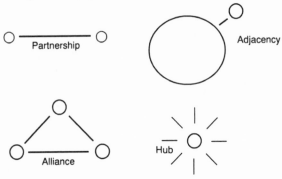

its retail trade. More on the foregoing can be found in Stover and Williams (1990).

Adjacencies. Sunbury, Ohio, is a small town that depends in part for its development on its adjacency to Columbus (Williams, Sawhney, and Brackenridge, 1990). Other examples have included Gibsonia near Pittsburgh, Pennsylvania; Catham near New York City; Dahlonega near Atlanta, Georgia; and Eagle Pass, Texas, across the border from Piedras Negras, Coahuila, Mexico (Schmandt et al., 1991; Strover and Williams, 1990). Although most of the examples we have studied show positive strategies for developing spatial relations, Donald Dillman (1985) has warned that telecommunications can also take business away from a small town, as in having accounting done "on-line," sending off typesetting, or using remote banking services from a larger town or city.

Alliances. This is where several small towns join together to promote development. None is a hub, but each may concentrate in certain areas of a development—for example, recreation, a community college, a library, a mall, industrial park, consolidated high school, a community college, or trade school. One example studied included the cluster of the small eastern Ohio towns of Lake Milton, Jackson, and Craig Beach. (Williams, Sawhney, and Brackenridge, 1990).

The implication is that telecommunications should be able to facilitate development within these spatial patterns. This includes having extended-area dial service (nontoll) within the developmental area, deploying networks capable of serving high-speed data and video-conferencing needs, and eventually seeing these areas as concentrations for intelligent network services.

RECOMMENDATIONS FOR RURAL
TELECOMMUNICATIONS IN AMERICA

On the Need for a Rural Policy

Although telecommunications policy research for rural America is still very much in an "in progress" state, we feel that a number of useful generalizations can be drawn at this point. These are made from the materials reviewed in this chapter and several ongoing studies.[6]

As we said in the introduction to this chapter, the status of the rural economy as well as the changing technological picture are driving the new interest in telecommunications and development in this country. The move away from seeing rural development as primarily agricultural, the coming of the intelligent network, and the increasingly flexible approach to regulatory change on both federal and state levels together present new opportunities for planning and innovation. The key problem is that rural telecommunications policy in America needs to be addressed in its own right, and not as an "appendix" to national or urban policy. This is the primary recommendation of this chapter, namely *that rural telecommunications applications for development be identified and promoted.* Some of the characteristics of that policy are as follows.

1. Define the Distinctive Needs of Rural Populations

We need to understand more about the distinctive needs of rural populations that can be served by new or improved telecommunications. On a more theoretical and abstract level, this will typically mean overcoming distance and lack of population density for achieving economies of scale for business or public service undertakings. Given an understanding of these needs, we should create new incentives in our telecommunications planning and regulatory structure to facilitate rural development. This eventually means promoting access to intelligent network services.

2. Improve Our Knowledge of Telecommunications and Rural Development

We need to understand more about the complementarity of rural economic development and the uses of telecommunications. To date, most of this evidence is descriptive and anecdotal, which provides some basis for interpretation. But it will be important to develop quantitative studies that identify relations between telecommunications investments and developmental outcomes. Moreover, researchers in telecommunications and researchers in rural development should improve their ability to work together. How can infrastructure development attain new economies of scale (e.g., an advanced network serving a region's business, residential, and public service needs)?

Among other strategies for action, this means that we should see how telecommunications can assist in promoting alliances, partnerships, or adjacency relationships in the development of rural areas. "Hub" patterns appear to be an important development strategy for rural areas, and telecommunications can facilitate such development. Networking of county public services, branch outlets of businesses, health clinics, and distance learning can all benefit from telecommunications services of this type. This can also be reflected in the "magnet" and "piggybacking" strategies where resources are consolidated to gain greater coverage. Leadership can be consolidated and coordinated in such patterns. Four towns working in an alliance, or four working around a hub will have greater chances for developmental success than if each operates as an isolate. The intelligent network infrastructure should be used to support such coordination.

Trials or demonstrations of services are an important strategy for diffusing new technologies, including uses of telecommunications. Developers should promote demonstration or development projects, and evaluate results. These exist today in distance education, but we should promote them in city administration, rural health care, and small business development. Telcos should be given special incentives to mount new demonstration projects that can be evaluated in quasi-experimental designs.

3. Evaluate Current Infrastructure and Set Priorities for Upgrades

The priorities developed from examination of relations between telecommunications investment and rural development should be a basis for defining "how we get there from here." First we need to know more about the status of current telecommunications services to the rural population, and get them up to date within the context of contemporary networks. As summarized earlier in Table 11.1, this includes reaching remote customers, single-party service, and digital switching, for the most part. The next priority is to set the course for advanced network services with an eye toward economic development such as described in the present set of recommendations. Finally, we should examine a longer-range view of deploying the intelligent network.

Telecommunications providers have a role that extends beyond providing or planning new services. They should take the lead in educating the public and decision-making officials as to the applications and benefits of bringing the intelligent network to the countryside. Hopefully, they will market more actively than they now do if conditions are created that make rural area investment attractive. If we are left to depend upon simply passive fulfilling of rural service requests, the rural environment will be left behind in the information age.

4. Determine New Strategies for Financing Rural Telecommunications

Somebody has to pay for these innovations, and we already know that many of the current shortcomings are due to lack of an attractive return on investment from rural telecommunications. Because these services cost more to install than in urban areas, we must anticipate new financing strategies. Traditionally, this has been in the form of low-interest loans and cross-subsidies. But to build a new network may require outright federal or state grants, perhaps along the lines of the interstate highway program of a near half century ago. Toward this end, and as advocated in the Aspen conference on rural telecommunications (Bollier, 1988), we should review models that in the past have contributed to development of the rural infrastructure—for example, the Tennessee Valley Authority (TVA), REA, and railroad and highway construction programs. It might be possible to see an expanded REA program for development of a rural area's entire network as a platform for integrated (i.e., business, education, health, residential) services.

5. Develop a New Rural Telecommunications Policy

We need to review critically existing U.S. telecommunications policy that either overlooks, or is at odds with, rural interests. It is likely that the definition of universal service will be upgraded to include access to data services and even particular types of "citizen content" (Chapter 12). It is important that we avoid the discrepancies or conflicts between urban and rural policies such as discussed earlier in this chapter. The shift from the regulatory to the competitive paradigm must be sure to take rural development into account, and not be biased, as usual, toward the cities. In the main, we are talking about *national* development that includes an effective urban-rural synergy. This national plan could encourage detailed actions on the state level where economic development programs and telecommunications planning could be carefully coordinated. Again, this means better liaison between state PUCs and development agencies.

Rural plans should encourage regulatory and market conditions for service integration. Just as there are economies of scale in expanding the scope of the network to include more customers, the same could be said of services. We should consider allowing rural service providers to expand their lines of business in rural areas, for example, into newly developed customer premises equipment, information, and video services. If a high-capacity network is built in a rural area, it should serve these multiple lines of business.

Finally, whatever initiatives are taken, we must be sure to balance the interests of large and small telecommunications companies. There are many rural areas where the local small telco or cooperative takes a major hand in development activities. In short, we must support the role of the small telco.

6. Encourage Active Developmental Roles for Service Providers

Service providers to rural areas should be encouraged to develop and disseminate company policies on rural services and their intended role in economic development. Telecommunications providers can help in meeting community needs in public service and education areas. There are developing examples of telcos teaming up with education or development agencies to develop demonstrations of distance learning (as by the Mid-Rivers Company in Glendive, Montana, or by the Minnesota distance education leaders in that state). Telecommunications providers have lent expertise in other areas as well, for example, in health care networks (see the Geisinger example in Chapter 8).

Rural telcos can attempt to integrate developmental objectives into state regulatory activities, including associating those objectives with state activities in economic development, as, for example, in activities of commerce departments or developmental committees. (There remains much development activity in such groups that is conducted in the absence of valuable information about uses of telecommunications.)

The same providers can also undertake special programs to inform their customers of ways that telecommunications can benefit their businesses. Telephone companies must develop a new mindset: they are selling more than just dial tones to rural businesses. They are selling competitiveness in the increasingly global, information-age economy.

Finally, rural service providers can themselves promote local economic development efforts. Southwestern Bell, for example, has reprinted several "economic development resource books" for distribution to their rural customers and local developmental groups.[7] Rural telecommunications companies know the local economy and could themselves in many cases be the leaders of developmental initiatives along the lines discussed in this report.

■   ■   ■

Relatively speaking, perhaps no geographic entity has more potential for benefiting from telecommunications than rural areas. Here the traditional problems of distance and lack of population density can, for many applications, be overcome or visibly reduced by applications of telecommunications. Rural development should be among the goals in the deployment of the intelligent network.

■

FIVE

The Future of the Intelligent Network

No part of the future of the intelligent network is more important than our human goals for its application. In most of the chapters of this volume we have seen examples of what business, residential customers, and public service agencies need and want in modern telecommunications services. Which of these do citizens deserve as basic—that is, *universal*—services? Given the coming intelligent network, what should be the new definition of universal service?

12

■

On the Redefinition
of Universal Service

*The position of the Bell System is well known. . . . The telephone
system should be universal, interdependent and intercommunicating,
affording opportunity for any subscriber of any exchange to
communicate with any other subscriber of any other exchange . . .
annihilating time or distance by use of electrical transmission.*

—Theodore Vail, AT&T ANNUAL REPORT (1910)

*. . . in light of the possibilities for new service offerings by the 21st
century, as well as the growing importance of telecommunications and
information services to U.S. economic and social development, limiting
our concept of universal service to the narrow provisions of basic voice
telephone service no longer serves the public interest. Added to universal
basic telephone service should be the broader concept of universal
opportunity to access these new technologies and applications.*

—National Telecommunications and Information Administration,
TELECOM 2000 (1988)

■　■　■

COMPETITION VS. UNIVERSAL SERVICE

No one can doubt that now in the over half decade since the divestiture
of AT&T, we are well on the road to a competitive marketplace in
telecommunications. Long distance is well on its way to being deregulated.

There is even abundant evidence of competition growing in the local exchange market, the very sector that many have argued would forever be a "natural monopoly." In all markets of this rapidly deregulating and increasingly competitive era, we will see the development of the intelligent network. But a key question is whether market forces alone (along with steady easing of the regulatory forces) can best serve our nation's telecommunications needs. Cost equalization is, and has long been, a fundamental issue. As discussed subsequently in this chapter, what will replace the equalization of rural versus urban services, residential versus business, or small telcos versus the Bell companies and large independents?

Further, how can we adjust to expanding services beyond basic voice— that is, to data, text, transaction, telemetry, and even video services? And with the coming of information services, how will we cope with the likelihood of wanting to specify certain types of content as basic service—as has occurred with 911 emergency calling, 976 messages, caller identification, and directory assistance? Will links with police, fire departments, schools, and even city hall become a basic service, or information such as hazardous warnings, governmental bulletins, the weather, traffic conditions, or medical advice?

All of the foregoing portend an increasing attention to the definition of basic telecommunications services due our citizens, in all likelihood an expanded definition of universal service. It is this policy concern that can balance citizen, business, and public service goals in deployment of the intelligent network as we move fully into the newly competitive environment. Accordingly, we should improve our grasp on this concept, and that is the goal in the materials that follow.

DEFINING UNIVERSAL SERVICE

An Agreed-Upon but Inexact Tradition

Perhaps no other regulatory goal has been so extensively discussed without an established definition.[1] The specific term "universal service" appears in no public law and there is no authoritative source defining precisely what it means, let alone how it might best be achieved (Gordon and Haring, 1988). Many agree that the basic goals of universal service are articulated in Article 1 of the Communications Act of 1934, requiring regulation to make available, so far as possible, to all people of the United States a rapid, efficient, nationwide, and worldwide wire and radio communication service with adequate facilities at reasonable charges.[2]

Today, universal service can mean many things depending upon the particular vantage point—public advocacy, economic, social, or legal—one chooses to use. Each of these offers a different rationale for universal service. For example, viewing universal service as a consumer or public advocacy issue, people may argue that it is a means of assuring that urban, rural, wealthy, and poor have the same access to information. Accordingly, regulators must assure that an "information elite" does not form in our society. Especially today, there is a danger that a small element within our society will have exclusive access to increasingly valuable information services as they become available (Pressler and Schieffer, 1988). This includes not only economic classes but urban rural issues (Brunner, 1986).

Public advocates may also argue that telephone and radio are indispensable lifesaving tools. Both broadcast media and telephone services such as 911 serve a necessary role in society. Viewed from an economic standpoint, the goal of universal service is important for several reasons. First is the concept of externality. The best phone is worthless if it is not hooked up to others. The greater the subscriber penetration, the more valuable the system is as a whole. Second, information networks are now considered part of our country's basic infrastructure. Without universal service, as can be interpreted from many application examples in this volume, it becomes more difficult for our country to remain competitive.

Information has also become much more important socially and culturally. Television and radio not only disseminate news and information, they serve an important cultural role. Many in Congress believe that the goals of universal service should extend to cable television and have introduced legislation to establish this goal.

Since universal service is not stated explicitly in any law, it is hard to defend legally. Many believe the importance of universal service warrants a more explicit legal definition than that found in the Communications Act.

Some Historical Notes

The concept of equal access fed concerns over telephone deployment as the industry grew in the late nineteenth century. Several states regulated telephone companies like public utilities as early as 1879. Regulation was seen as a means to consolidate the industry and avoid wasteful and inefficient resource allocation. These early state and local regulations contain the first articulations of the goal of universal service. By 1910, Theodore Vail, then chairman of AT&T, first promoted the idea of national universal telephone service in the company's Annual Report (quoted in the chapter opening). To Vail, universal service meant promoting citizen access by linking up the nation's proliferating but isolated networks (see Dordick, 1990, for a historical view on this). He hoped to achieve this goal by supporting state regulation of the industry. Because economies of scale are inherent in the technology, he believed that a regulated monopoly would be the only way to achieve his goal.

This concept became law as it was reflected in the Communications Act of 1934. The telecommunications industry grew rapidly in the ensuing decades. But as a national network began to emerge, state regulators found it increasingly difficult to deal with AT&T and its growing monopoly. In 1930, AT&T served 80 percent of local exchange customers; its subsidiary, Western Electric, sold 92 percent of all telecommunications equipment; and AT&T Long Lines provided 100 percent of all long-distance service. Despite its overwhelming dominance of the industry, federal regulation was virtually nonexistent at the time.

As a result, Congress commissioned a study of the telecommunications industry. The recommendations of this study resulted in the Communications Act of 1934. When Congress passed the act, they consolidated control over virtually all forms of communication into one body, the Federal Communications Commission (FCC). The Communications Act empowers the FCC to fix rates and control accounting practices of interstate telephone and telegraph companies. The act also authorizes the FCC to issue licenses and assign frequency bands to those in the field of wireless communication. Not only did Congress willingly yield its own power and vest it in the FCC, it gave the FCC wide discretion in its decisions over the communications industry.

The Communications Act's most significant impact may have been its statement of purpose, Article 1 of the Communications Act of 1934. Its social objective significantly impacted the way costs would be allocated and rates would be set. Indeed, separating facilities and allocating costs has been one of the commission's primary tasks since its inception.

As the national telecommunications network grew, state regulators found it increasingly difficult to coordinate their regulation. For rates to be "just and reasonable," it would be necessary to allocate costs and separate the rate base between state and federal jurisdictions. But given the interconnectedness of the network, any separation of it would be somewhat arbitrary. For example, to make a long-distance call, one would use the local exchange plant at each end as well as the toll facilities.

To separate and price these elements consistently, some rationale would have to be developed. Before 1930, state regulators relied upon a "board-to-board" approach for pricing telephone service. According to this philosophy, all of the costs associated with the local exchange were recovered from local rates. The rationale behind this philosophy was largely practical, as calculating any kind of separation of the network would be very difficult.

But in 1930, the Supreme Court decided that local rates were too high because all of the fixed, nontraffic-sensitive (NTS) costs were allocated to intrastate service. In *Smith v. Illinois Bell*, the Supreme Court ruled that some of the NTS costs be allocated to long-distance service. While the court recognized the difficulty in separating the local plant between state and federal use, it recognized that without knowing detailed uses of the services this could become a burden to intrastate rates.

This decision resulted in the eventual structure of the U.S. telecommunications industry—pooling and nationwide rate-averaging—that has enabled most Americans access to affordable basic telephone service. Further revisions in separations and costing were required as advances in technology allowed the cost of long-distance service to decline more rapidly than that of local service. The cost allocation formula was continually adjusted in favor of the local exchange. In addition, the FCC adopted a policy of "equal charges for equal service" in 1941 to reduce interstate rate discrepancies.

The 1948 Rural Electrification Act, which established the Rural Electrification Administration (REA), further facilitated the goals of universal service. The REA has enabled rural areas access to telephone service by making low-cost loans to rural telcos to build and expand local networks (see Chapter 11).

All of these policies and programs contributed to meeting the goals of universal service. In the early fifties, with AT&T operating almost entirely under a nationwide average pricing system, almost 80 percent of American homes had at least one telephone. Within this giant regulated monopoly it was possible to have a structure of cross-subsidies. Rate-averaging for rural customers could be supported by urban funds. Funds from long distance and business-use funds could keep residential rates low. And independent companies could benefit from their political support of the system. (See Faulhaber, 1987, for more on this.)

As technology advanced and more competitors entered into the telecommunications market, this system became harder to maintain. Competitors were able to set prices of equipment and services closer to costs. As a result, AT&T reacted strongly to the slightest competition in the industry. AT&T was able to hold on to its monopoly up until the late 1960s, but by then the distortions of its monopolized services deviated too far from the realities of economic and technological change. The onset of deregulation and divestiture changed the game.

TOWARD REDEFINITION

Implications of Deregulation

Because the initial goals of universal service—affordable basic voice telephone service—have largely been met, many question the purpose of the old regulatory structures. These people believe that regulators need only ensure that everyone can continue to afford basic telephone service. They believe that this can be accomplished by providing direct subsidies, like the lifeline concept, to those who cannot afford service.

Lifeline was introduced in several regulatory bodies as a way of preserving the goals of universal service in light of the general movement toward cost-based pricing. Lifeline telephone service offers a limited number of local calls at a price below cost for residential subscribers who meet an established means test. The FCC has implemented two optional lifeline programs: one is a matching fund for state lifeline efforts; the other, called Link-Up America, is designed to offset installation costs. Congress has attempted to extend the federal lifeline program for the past several years but its efforts have not yet succeeded. States have also developed lifeline programs and the FCC voted to expand its voluntary program. However, the uneven nature of the states' participation and the limited nature of the FCC's voluntary program prompt lifeline advocates to continue pressing for federal legislation.

The original idea of a lifeline emerged from the FCC's 1982 Access Charge Order. The implementation of the subscriber line charge raised concerns that some subscribers would no longer be able to afford telephone services. The FCC maintained its support for shifting NTS costs from interstate users to all subscribers via a flat monthly charge on all users. However, it stated its willingness to "entertain waiver requests" of its residential subscriber line charge from local telephone carriers who wished to provide lifeline options to preserve universal service.

Thus, a lifeline program was developed to offset the subscriber line charge, despite the FCC and Joint Board's assertion that "the proposed subscriber line charges should not have an adverse effect on universal service" (FCC, 1985). The program is voluntary and state lifeline programs must be established before certification in the FCC program is granted. The amount of federal assistance depends upon how much assistance is provided by the states.

In 1987, the Joint Board recommended that the FCC enhance its lifeline program with a federally funded "Link-Up America" plan to assist subscribers in paying for initial subscriber hookup costs. The FCC established the program and now contributes up to 30 dollars per eligible subscriber to offset up to half of the charge for initiating service. In addition, it assists with the remaining amount by paying the interest charged by the local telephone company.

Controversy over the lifeline program has emerged not because people oppose it—almost everyone believes that some mechanism should be in place to promote universal service—but because of the manner in which it should be accomplished. Some of the issues debated are: the desirability of a government mandated versus a voluntary marketplace approach; the design of program offerings; and the responsibility for program administration and implementation.

The United States Telephone Association (USTA) criticized the financing of the programs through long-distance charges (and ultimately ratepayers). Consumer groups have also criticized the requirement that limits application for federal lifeline assistance solely to state PUCs and local exchange carriers. These groups believe that consumers and their "direct representatives" should be able to apply directly to these programs.

Congress has tried in the past to address some of these concerns with federal lifeline legislation. Legislation was introduced in the 98th, 99th, and 100th Congresses to no avail. Consumer, labor, religious, and senior citizens groups support legislation for nationwide, mandatory lifeline service. These groups argue that basic telephone service is a necessity, not a luxury. Opponents of federally mandated lifeline programs feel that the market or that the states themselves should be responsible for implementation of programs.

Views on Expanding Universal Service

Some reject the idea that universal service has been accomplished. These people believe that universal service is a relative term whose meaning should change as technology advances. Because information plays such a paramount role in our society today, people adhering to this view feel that the original goals of universal service are no longer sufficient. According to this view, regulators need to not only assure continued low rates, but need to adopt policies to ensure that our society does not become one of information haves and have-nots.

It may be necessary to redefine what constitutes basic service as our economy changes. For example, the service sector is the fastest-growing sector in both rural and urban America today. Information has become a critical resourse in not only the service sector, but even in more traditional industries such as agriculture and manufacturing. Technology can also provide quality education and health care to areas that lack these essential services.

The Rural Telephone Coalition, consisting of the National Telephone Cooperative Association (NTCA), the National Rural Telecom Association (NRTA), and the Organization for the Protection and Advancement of Small Telephone Companies (OPASTCO), recommended that Congress establish *universal information service* and a nationwide intelligent network as national goals. They also recommend that Congress make use of the federal-state Joint Board process to evaluate universal service needs. The United States Telephone Association (USTA) had made similar recommendations.

While many may agree that either an implicit or explicit redefinition of univeral service is necessary, conflicts arise regarding the correct way to achieve this goal. This becomes very clear at the state level of regulation and decision making. There are three general positions. The first is that increased competition undermines the goals of universal service and that competition should emerge only gradually. These people believe that the old monopoly system was the best model to achieve reasonably priced basic service. A second position sees that competition will eventually bring better technology and services to local subscribers. These people believe in increased competition and deregulation and greater state control of regulation. The last position involves a much more radical belief that competition should even be allowed at the local level to "eliminate the perverse incentives of regulated monopoly" (Noll, 1988, p. 6).

The National Telecommunications and Information Administration (NTIA) believes that less, rather than more regulation in the industry will remove the barriers to expanded access. They attribute increasing usage of bypass to "inefficient and unwarranted regulation." As bypass increases, the public network has less reason to make investments in technology and service offerings. Since large users usually require the most advanced capabilities, it is their demand which determines how advanced a telephone network will be. The NTIA also believes that legal obstacles, such as the line-of-business restrictions in the Modified Final Judgment, impede investment. Although the NTIA supports programs such as lifeline for the poor and rate-averaging for high-cost areas, they believe that efforts to keep rates down artificially—such as lengthy depreciation schedules, cross-subsidies from long distance, business users, and vertical services—are only beneficial in the short run. Furthermore, these may have devastating long-run consequences. In particular, the NTIA maintains that the continued use of traditional cross-subsidies in the industry will delay movement into "the information age."

Although the NTIA is correct in stating the need to redefine universal service, its view that deregulation will achieve this goal could be flawed. For example, rural areas, as described in Chapter 11, simply do not have the economies of scale to build and operate the network with prices at costs. (Providing telephone service to rural areas can cost ten times more than in urban ones.)

Michael Brunner (1986), executive vice president of the National Telephone Cooperative Association, believes that federal regulators and Congress have, for the most part, fulfilled their "obligation" of protecting and preserving universal service. He believes that the real threat to universal service lies in many states' willingness to dispose of the mechanisms that have thus far preserved universal service. For example, the New Mexico public utility commission recently adopted an originating responsibility plan which will eventually require each local company to calculate and assess the cost of toll calls that its subscribers originate. This is, in essence, toll rate de-averaging which puts rural residents at a great disadvantage.

The OTA, in *Critical Connections* (1990), offers several strategies and options Congress could pursue to implement universal service. These include restructuring the prices at which communication services are offered, providing direct government support for users to access information and communication paths, and assuming a more proactive role in assuring lively debate on issues of public importance.

One of the most visible reports holding implications for redefinition of universal service was the California Intelligent Network Task Force, discussed next.

California's "Intelligent Network Task Force"

Although Pacific Bell sponsored its developments and meetings, the Intelligent Network Task Force in California had all of the characteristics of a consumer advocacy group. Chaired by Dr. Barbara O'Connor, a Sacramento professor long experienced in public uses of media, the Task Force both promoted study by, and solicited subsequent suggestions from, representatives of many of the subgroups of California society. The group met monthly over 18 months to study options and to develop priorities for the applications of telecommunications for the advancement of the citizens of the state. They also introduced themselves to many of the new services: group members learned to use personal computers for access to database services, electronic mail, and report writing.

In the eyes of this Task Force, the "intelligent network" is a public network offering a range of advanced voice and data services to the public. Contrary to the views of some consumer advocates who hold that the

Table *12.1.* California Intelligent Network Task Force Recommendations

1. Those responsible for telecommunications regulation should authorize, and Pacific Bell should build and make available to the people of California, an intelligent network that incorporates the qualities envisioned in the Task Force's definition.

2. Universal service should be redefined by regulators and the telecommunications industry to include access to the intelligent network and to a specific set of essential applications services, as envisioned in the Task Force definition.

3. In order for the benefits of the information age to be available to virtually all Californians in the most efficient and economical way, Pacific Bell must design intelligence into the network and incorporate the most important elements of the system into a redefined universal service.

4. The California Public Utilities Commission should form a joint advisory committee of Commission staff, Pacific Bell staff, and consumer representatives to discuss rates, regulation, and cost subsidy issues relative to development of an intelligent network.

5. Telecommunications companies, regulatory agencies, consumers and legislators, working together, should devise appropriate standards and techniques to assure system security and protect user privacy. It is recognized there are difficult issues and significant costs, but in a democracy few things are ultimately more important than the personal privacy of the individual.

6. The intelligent network has the potential to change the ways in which a substantial amount of the society's work is done. Pacific Bell, unions, and other appropriate interest groups should study jointly the implications of such changing work environments as telecommuting, in order to maximize their undoubted advantages, while avoiding a range of potentially serious pitfalls and abuses such as social isolation, promotability, plus others.

7. Pacific Bell and the leaders of public education should jointly plan applications of the Intelligent Network to enhance education in schools and colleges, in the workplace, and at home. An education task force should be convened to advise and help plan applications for intelligent network services. In the Information Age, a rising standard of excellence is required in education. There must be an educational continuum that begins in kindergarten and extends through one's working life and beyond. The intelligent network will be integral to that continuum.

8. Pacific Bell should expand its expertise in instructional telecommunications applications. A separate entity dedicated to educational applications and market development should be formed, utilizing staff with professional experience and training in education.

9. Pacific Bell, other information technology companies, and health care providers should work jointly for the further development of in-home health care monitoring facilities, emergency call systems, and related services.

Table *12.1.* California Intelligent Network Task Force Recommendations
(con't)

10. Pacific Bell should assure, as the intelligent network evolves, that there are suitable provisions for access by those whose disabilities preclude use of "conventional" facilities. The company should actively seek ways in which the network can enhance the ability of disabled persons to compete in the job market and otherwise participate fully in the life of the community.

11. Pacific Bell, and others in the industry, should press ahead with development of automatic language translation and other techniques to facilitate use of the intelligent network by those not fluent in English.

12. The Intelligent Network Task Force suggests that it remain in place to meet with Pacific Bell on a semi-annual basis to discuss progress on the recommendations. Additionally, after extensive first-hand experience and discussion of information-age issues, we can be effective consumer advocates before political and regulatory bodies and can facilitate partnerships between the groups we represent.

Courtesy of Dr. Barbara O'Connor. Pacific Bell distributes copies of this report under the title *Pacific Bell's Response to the Intelligent Network Task Force Report* (1988).

general public only wants inexpensive "plain old telephone service," the California group advocated that universal service definitions be upgraded to reflect the services available from an advanced network. They further advocated that a joint advisory committee of industry and public representatives develop a plan for financing the network.

Their recommendations represent a citizen-based bid for an advanced state telecommunications environment, not unlike California's successful initiatives in water, transportation, and education. The recommendations of the Task Force are summarized in Table 12.1. Subsequent to its circulation in interim report form, Pacific Bell (c. 1988) published a "glossy" edition featuring its positive responses to the suggestion. Since that publication, regulatory conditions have moved in California toward making the building of that network a possibility. It remains to be seen whether Pacific Bell will take up the challenge it invited.

THE POLICY CHALLENGE

As we said in the opening chapter of this book, the intelligent network, with its greatly enhanced services, will increase our economic and social options. Whereas already in Japan and over the next few years in the unified European Common Market, the intelligent network is a national or regional policy, its continued development remains a topic of debate in the

United States. Despite the existence of visions like *Telecom 2000* (NTIA, 1988) or *Critical Connections* (OTA, 1990), most of the activity in telecommuncations policy is still in the transition stage of being reactive to divestiture rather than being proactive in plotting our future. We are devoting large amounts of attention in moving from regulation to competition, but not enough to the personal, business, or developmental applications of telecommunications.

In postdivestiture times, as summarized in Figure 12.1, we have treated universal service, or "threats thereto," as more of a negative concept—a "warning" signal—than a positive planning goal. In the early years of the public network, and even in the post–World War II years as telephone penetration passed the 90 percent mark, basic "connectivity" was a positive policy concept. But now we tend to treat it more negatively where threats to same are a more or less counterbalance to the move to competition. In so doing, we have made too much of competition being an end in itself in current policy making. Yes, we do consider competition as a means to stimulate development of the network, but it is not a suitable policy end or goal. Perhaps it is now time, as also illustrated in Figure 12.1, to recommit policy goals to universal service—a redefined version that takes into account the new capabilities of the network. We can still retain a commitment to competition but as a means not an end.

Based on the type of "usage" or "demand-side" research cited in this volume, we hold that there are many arguments for business, public service, and residential uses of the intelligent network. We have many new bases for redefining universal service, and we should promote debate in that direction. The foregoing leads into the major issue of what the expansion of data or information services and the coming of video services will do to the traditional understanding that universal service—a policy premise of the public network—is access to basic voice communications. Will voice remain the basic inexpensive and widely available service, while all else is in the premium category? This could lead to a two-tiered network, where that part of the population who might be in most need of the new services (emergency, education, information) will be the least able to afford them. Perhaps universal service should involve some level of access to multimedia services, and possibly some types of vital information (e.g., weather, governmental announcements, hazardous warnings; see Williams and Hadden, 1991).

As we have written in earlier chapters, computing and telecommunications are coalescing into a network capable not only of high-capacity transmission of voice, data, and video services, but also of being a growing information resource. At the same time, structural changes in the economy

Figure *12.1.* Redefinition and Repositioning Universal Service as a Policy Goal

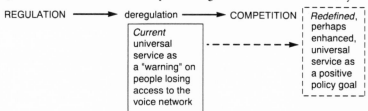

and the growth of international markets have increased the demand for more advanced and innovative information processing and transmission. These forces are creating pressures to further deregulate the telecommunications industry, as most believe that competition is the best means of meeting these new needs. But how can we nurture the delicate balance between letting the market lead us, as against ensuring availability of certain services vital to a citizen's life and work in the information age? A renewed definition of and a commitment to universal service as the overarching policy goal in national telecommunications is our critical challenge in planning for the coming intelligence network.

■   ■   ■

In all, the coming of the intelligent network may be as much or more a case of facing up to telecommunications policy challenges of what we wish to set as universal service goals in the United States, as it is a case of simply "regulation to competition transition," engineering breakthrough, or new market visions.

Notes

Chapter 4. Networking Large Businesses

1. From an original report by Liching Sung, in Schmandt et al. (1991). The research involved visits to Wal-Mart facilities, including the headquarters in Bentonville, interviews with key personnel, especially Jay Allen and Tom Newell at headquarters. Other sources of information, including published ones, are cited in the original report.

Chapter 5. Doing Business on a Smaller Scale

1. From Strover and Williams, 1990; Joan Stuller (March 8, 1990) interview with Shirley Jay, owner, Century 21 Realty, Demopolis, Alabama.
2. Scott Klopfenstein, Prodeva Corp. (Williams, Sawhney, and Brackenridge, 1990).
3. Richard Cutler interviews with Wil Bocum, Maverick County Agricultural Extension Agent, Eagle Pass, Texas, and Ben Doherty, Sales Representative, Alta Verde Industries, Inc., Quemado, Texas; summer 1990 (Schmandt et al., 1991).
4. From our personal experience in talking with telemarketers nationally, the relatively "neutral" midwestern accent may not be as important as basic intelligibility. Some small businesses in the deep South or Texas using telemarketing feel they have no problems with attitudes about regional accents unless it is difficult for others to understand the person. In fact, if you are selling cowboy boots, a Texas accent might be of benefit.

Chapter 6. Promoting Science and Technology Transfer

1. For more on Tecnopolis Novus Ortus, see Bozzo and Gibson (1990).
2. Many of the network examples in this chapter are drawn from research by Eloise Brackenridge; see Williams and Brackenridge (1990).

Chapter 7. Making a Residence a Home

1. Prospects for this type of business was the topic of a multiple client research project reported in Dordick, Bradley, and Nanus (1980). For a later view of the publisher-sponsored text services, see Greenberger (1985).

Chapter 8. Bringing Innovation to Public Services

1. Much of the material in this section is drawn from Schmandt et al. (1991), Chapter 3, especially the research by Paula Adams. More details, interview sources, and references can be found in that material.
2. This example is from Strover and Williams (1990); field research by Liching Sung.
3. From Schmandt, Williams, Wilson, and Strover (1991); field research by Scott Lewis.
4. From Williams, Sawhney, and Brackenridge (1990); field research by Harmeet Sawhney.
5. From Schmandt et al. (1991); field research by Robert Stephens.
6. From Schmandt, Williams, and Wilson (1989); see Chapter 3 by Darrick Eugene and David McCarty.
7. From Schmandt, Williams, and Wilson (1989); see Chapter 2 by Danny B. Garcia and Mahmoud Watad.

Chapter 9. States: New Roles and Initiatives

1. The Williams (1988a) report gives more details of these results.

Chapter 10. Cities: New Urban Infrastructures

1. We studied Miami both in the state and city studies, and the author has visited and spoken at LATCOM. This material was gathered by Jewell Evans and appears in Chapter 7 of Schmandt et al. (1990).
2. From materials by Larkin Jennings and Harmeet Sawhney, Chapter 6 in Schmandt, Williams, and Wilson (1989).
3. From materials by Danny B. Garcia and Mahmoud Watad, Chapter 2 in Schmandt, Williams, and Wilson (1989).

Chapter 11. Rural: Developing the Land Between the Cities

1. The XIT telephone cooperative was a study site in Schmandt et al. (1991), visited by Harmeet Sawhney and Frederick Williams.
2. The statistics were drawn from their "Rural Economic Development in the 1980s: A Summary, *Agricultural Information Bulletin*, Number 533, Oct.

1987, and Thomas F. Hady and Peggy J. Ross, *An Update: The Diverse Social and Economic Structure of Nonmetropolitan America* (AGES 9036), May 1990.

3. This summary was first distributed as an Aspen report (Bollier, 1988) and later reproduced as an appendix in Parker et al. (1989).

4. The latter two organizations may be less known; Cabela's is a retailer of outdoor recreation items; EMERG provides research and data processing services to college financial aid offices.

5. One, a study of rural sites in Ohio (Williams, Sawhney, and Brackenridge, 1990) involved studies of small business, residential, and public service needs. The second (Strover and Williams, 1990), a Ford Foundation sponsored project conducted in conjunction with the Aspen Institute Rural Poverty Program, involved in-depth studies of four small cities (Eagle Pass, Texas; Demopolis, Alabama; Kearney, Nebraska; Glendive, Montana). Although examples from these studies are cited in the present volume, their overall findings were not available at this time.

6. This includes our completion of the project inquiring into telecommunications of the life of four small cities (Strover and Williams, 1990), and other projects inquiring into rural telecommunications policy, including the work of Edwin Parker and Heather Hudson in analysis of state investments and planning. In another five or so years, we should have evidence to test the success of rural initiatives of the last several years.

7. The booklets are "A Rural Economic Development Source Book" and "Profiles in Rural Economic Development." Both were researched and produced originally by the Economic Development Administration, U.S. Department of Commerce, Washington, D.C., 20230.

Chapter 12. On the Redefinition of Universal Service

1. The background material on universal service has been drawn from a report by Jill Ehrlich, as commissioned by the author.

2. The language of Article 1 is analyzed phrase by phrase in Pressler and Schieffer (1988).

Glossary

access charge: A special fee to compensate the local exchange company for use of its network to connect to the long-distance network; a fixed fee for access has been authorized to be charged to U.S. telephone customers.

analog: Signal representations that bear some physical relationship to the original quantity; usually electrical voltage, frequency, resistance, or mechanical translation or rotation.

analog switch: A telecommunications switch that operates with analog signals (rather than digital).

antenna: A device used to collect or radiate radio energy.

bandwidth: The width of an electrical transmission path or circuit, in terms of the range of frequencies it can pass; a measure of the volume of communications traffic that the channel can carry. A voice channel typically has a bandwidth of 4,000 cycles per second; a TV channel requires about 6.5 MHz.

baseband: An information or message signal whose content extends from a frequency near dc to some finite value. For voice, baseband extends from 300 hertz (Hz) to 3400 Hz. Video baseband is from 50 Hz to 4.2 MHz.

baud: Speed of transmission in bits per second (bps) in a binary (two-state) telecommunications transmission. After Émile Baudot, the inventor of the asynchronous telegraph printer.

binary: A numbering system having only digits, typically 0 and 1.

bit: Binary digit; the smallest part of information with values or states of 0 or 1, or yes or no. In an electrical communication system, a bit can be represented by the presence or absence of a pulse.

BOC: Telephone jargon for "Bell operating company," used to refer to divested companies.

broadband carriers: The term to describe high-capacity transmission systems used to carry large blocks of telephone channels or one or more video channels. Such broadband systems may be provided by coaxial cables, microwave radio systems, or optical fibers.

broadband communication: A communication system with a bandwidth greater than voiceband. Cable is a broadband communication system with a bandwidth usually from 5 MHz to 450 MHz.

bypass: A telephone industry term meaning service that avoids use of the local exchange company network, such as a customer connecting directly into the long-distance network or buying a direct line between offices instead of using the public network.

byte: A group of bits processed or operating together; 16-bit and 32-bit bytes are common.

CAD: Computer-aided design; techniques that use computers to help design machinery and electronic components.

CAE: Computer-assisted engineering.

CAI: Computer-assisted instruction.

CAM: Computer-aided manufacturing.

carrier: Signal with given frequency, amplitude, and phase characteristics that is modulated in order to transmit messages. A colloquial use can refer to a telecommunications company.

carrier signal: The tone that you hear when you manually dial into a computer network.

cathode-ray tube: Called CRT, this is the display unit or screen of a computer or terminal.

CCITT: Consultative Committee for International Telephone and Telegraph, an arm of the International Telecommunications Union (ITU), which establishes voluntary standards for telephone and telegraph interconnection.

cellular radio (telephone):   Radio or telephone system that operates within a grid of low-powered radio sender-receivers. As a user travels to different locations on the grid, different receiver-transmitters automatically support the message traffic. This is the basis for modern cellular telephone systems.

central office: The local switch for a telephone exchange, sometimes referred to as a "wire center."

channel: A segment of bandwidth that may be used to establish a communications link. A television channel has a bandwidth of 6 MHz, a voice channel about 4000 Hz.

chip: A single device made up of transistors, diodes, and other components, interconnected by chemical process and forming the basic component of microprocessors.

circuit switching: The process by which a physical interconnection is made between two circuits or channels.

coaxial cable: A metal cable consisting of a conductor surrounded by another conductor in the form of a tube that can carry broadband signals by guiding high-frequency electromagnetic radiation.

common carrier: An organization licensed by the Federal Communications Commission (FCC) and/or by various state public utility commissions to supply communications services to all users at established and stated prices.

compressed video: Televison systems that require much less bandwidth than commercial television standards; used in business and educational settings.

COMSAT: Communications Satellite Corporation. A private corporation authorized by the Communications Satellite Act of 1962 to represent the United States in international satellite communications and to operate domestic and international satellites.

CPE: Telephone jargon for "customer premises equipment" which may often be distinguished from telephone company-owned equipment.

cross-subsidy: In telecommunications, this means that funds from one part of the business (e.g., long distance) are used to lower prices in another (local service). A controversy is how to prevent cross-subsidy between regulated and unregulated parts of the telephone business.

CRT: See *cathode-ray tube.*

custom calling features: Services like call waiting, call forwarding, and speed dialing to which one can subscribe.

database: Information or files stored in a computer for subsequent retrieval and use. Many of the services obtained from information or videotex services involve accessing large databases.

de-averaging: Changing telephone rates so as to reflect true cost differences, thus making rates vary in different parts of a state. (Local rates are typically regulated so that telephone service is not much more expensive in some parts of a state than in others, although the costs to the providers may vary greatly; rates are kept at an "average" by having a pool so that high-cost areas are subsidized by low-cost ones. Typically, rural telephone companies are against de-averaging because it could cause a major increase in their rates.

dedicated lines: Telephone lines leased for a specific term between specific points on a network, usually to provide special services not otherwise available on the public switched network.

depreciation: As usually defined, the tax "write-off" or giving credit in some way for the declining value of equipment investments; in the telephone business, depreciation variations are an important variable in setting rates.

digital: A function that operates in discrete steps as contrasted with a continuous or analog function. Digital computers manipulate numbers encoded into binary (on-off) forms, while analog computers sum continuously varying forms. Digital communication is the transmission of information using discontinuous, discrete electrical or electromagnetic signals that change in frequency, polarity, or amplitude. Analog forms may be encoded for transmission on digital communication systems (see *pulse code modulation*).

digital switch: A telecommunications switch that operates with digital signals (rather than analog).

direct broadcast satellite (DBS): A satellite system designed with sufficient power so that inexpensive earth stations can be used for direct residential or community reception, thus reducing the need for local communications by allowing use of receiving antennae with a diameter of less than one meter.

divestiture: The break-up of AT&T (1984) into separate companies.

dominance: A telephone industry term meaning whether a company serving an area has such a high percentage of the business that it drives out competition and that it can set prices. A current challenge is in how to define and measure dominance.

downlink: An antenna designed to receive signals from a communications satellite (see *uplink*).

earth station: A communications station on the surface of the earth used to communicate with a satellite. (Also TVRO, television receive only earth station.)

electronic mail ("e-mail"):   The delivery of correspondence, including graphics, by electronic means, usually by the interconnection of computers, word processors, or facsimile equipment.

ESS: Electronic switching system; the Bell system designation for their stored program control switching machines.

expert system: Computer software which embodies human-like decision capabilities, typically developed from analysis of observed types of expertise in human behavior.

FAX: Facsimile; a system for the transmission of images. It is a black-and-white reproduction of a document or picture transmitted over a telephone or other transmission system.

FCC: Federal Communications Commission; a board of five members (commissioners) appointed by the President and confirmed by the Senate under the provision of the Communications Act of 1934. The FCC has the power to regulate interstate communications.

fiber optics: Glass strands that allow transmission of modulated light waves for communication.

final mile: The communications systems required to get from the earth station to where the information or program is to be received and used. Terrestrial broadcasting from local stations and/or cable television systems provide the final mile for today's satellite networks.

frequency: The number of recurrences of a phenomenon during a specified period of time. Electrical frequency is expressed in hertz, equivalent to cycles per second.

frequency spectrum: A term describing a range of frequencies of electromagnetic waves in radio terms; the range of frequencies useful for radio communication, from about 10 Hz to 3000 GHz.

gateway: The ability of one information service to transfer you to another one, as when you go from Dow Jones News Retrieval to MCI Mail.

geostationary satellite: A satellite, with a circular orbit 22,400 miles in space, which lies in the satellite plane of the Earth's equator and which turns about the polar axis of the Earth in the same direction and with the same period as that of the Earth's rotation. Thus, the satellite is stationary when viewed from the Earth.

gigahertz (GHz): Billion cycles per second.

hardware: The electrical and mechanical equipment used in telecommunications and computer systems (see *software*).

headend: The electronic control center of the cable television system where weaving signals are amplified, filtered, or converted as necessary. The headend is usually located at or near the antenna site.

hertz (Hz): The frequency of an electric or electromagnetic wave in cycles per second, named after Heinrich Hertz who detected such waves in 1883.

hypermedia: A nonlinear file and presentational mode of information whereby in a description of a given topic, the user could consult any word or image in the display and gain access to further files. For example, you might be reading a file on computer-assisted design (CAD); a hypermedia interface would allow you to point ("click") to any example in that file so as to furnish more details. Apple Computer's Hypercard is the best-known example of hypermedia. Presumably, this is a human-and-machine interface more akin to normal cognitive processes than using a branching menu or access codes for subfiles.

information utility: A term increasingly used to refer to services that offer a wide variety of information, communications, and computing services to subscribers; examples are The Source, CompuServe, or Dow Jones News Retrieval.

institutional loop (or "net," I-Net): A separate cable for a CATV (community antenna television) system designed to serve public institutions or businesses usually with two-way video and data services.

interface: Devices that operate at a common boundary of adjacent components or systems and that enable these components or systems to interchange information.

ISDN: Integrated services digital network; a set of standards for integrating voice, data, and image communication; a service now being promoted by AT&T and some regional telephone companies.

IXC: Interexchange carrier; telephone companies (e.g., AT&T, MCI, US Sprint) that connect local exchanges and local access and transport areas (LATAs) to one another; a highly competitive part of the business.

Joint Board: A board of four state commissioners and three FCC commissioners appointed to settle federal-state regulatory issues or disputes.

K: 1024 bytes of information, or roughly the same number of symbols, or digits.

kilohertz (KHz): Thousand cycles per second.

LAN: See *local area network*.

laser: Light amplification by simulated emission of radiation. An intense beam that can be modulated for communications.

LATA: Local access and transport area; a telephone service region incorporating local exchanges, yet usually smaller than a state; typically they are serviced by a given telephone company for local services, and interexchange carriers for some intraLATA and all interLATA service.

LEC: Local exchange company; the telephone company that supports local calls (non–long distance); typically a regulated monopoly. LECs are certificated for areas called LATAs (local access and transport areas).

LMS: Local measured service; a method of telephone rate calculation that is sensitive to amount of usage as against a flat rate.

local area network (LAN): A special linkage of computers or other communications devices into their own network for use by an individual or organization.

Local area networks are part of the modern trend of office communications systems.

loop: The link that extends from a telephone central office to a telephone instrument. The coaxial cable in broadband or CATV system that passes by each building or residence on a street and connects with the trunk cable at a neighborhood node is often called the "subscriber loop" or "local loop."

LSI: Large scale integration; applies to microchip design.

megahertz (MHz):   Million cycles per second.

memory: One of the basic components of a central processing unit (CPU). It stores information for future use.

MFJ: Short for "modified final judgment," an antitrust agreement, that set AT&T divestiture in motion.

microchip: An electronic circuit with multiple solid-state devices engraved through photolithographic or microbeam processes on one substrate.

microsecond: One-millionth of a second.

microwave: The short wave lengths from 1 GHz to 30 GHz used for radio, television, and satellite systems.

millisecond: One-thousandth of a second.

millisecond: One-thousandth of a second.

minicomputer: In general, a minicomputer is a stationary computer that has more computer power than a microcomputer but less than a large mainframe computer.

modem: Short for "modulator-demodulator." The equipment that you use to link your computer to a telephone line.

modulation: A process of modifying the characteristics of a propagating signal, such as a carrier, so that it represents the instantaneous changes of another signal. The carrier wave can change its amplitude (AM), its frequency (FM), its phase, or its duration (pulse code modulation), or combinations of these.

monitor (video): Usually refers to the video screen on a computer but has more technical meanings as well.

MOU: Minute of use; a usage measure used in the telephone business to calculate certain rates.

multiplexing: A process of combining two or more signals from separate sources into a single signal for sending on a transmission system from which the original signals may be recovered.

nanosecond: One-billionth of a second.

narrowband communication: A communication system capable of carrying only voice or relatively slow-speed computer signals.

network: The circuits over which computers or other devices may be connected with one another, such as over the telephone network; or one can speak of computer networking.

node: A point at which terminals and other computer and telecommunications equipment are connected to the transmission network.

N.T.I.A.: National Telecommunications and Information Administration (U.S. Department of Commerce).

off-line: Equipment not connected to a telecommunications system or an operating computer system.

ONA: See *open network architecture*.

on-line: Being actively connected to a network or computer system; usually being able interactively to exchange data, commands, information with a host device.

open network architecture: These are standards that allow different telecommunications vendors to interconnect with a network.

optical fiber: A thin flexible glass fiber the size of a human hair which will transmit light waves capable of carrying large amount of information.

PABX: See *PBX*.

packet switching: A technique of switching digital signals with computers wherein the signal stream is broken into packets and reassembled in the correct sequence at the destination.

PBX: A private branch exchange which may or may not be automated. Also called PABX (private automatic branch exchange).

PCN: See *personal communications network*

personal communications network: Currently under development, this is a short-range, low-power, digital radio link for voice and data terminals that can be accessed through one's personal user number. It provides for portability in local areas, but can be coded to "travel" with the user to different areas. Currently, these services have the prospect of being sold by organizations other than the local exchange company in U.S. markets, an obvious policy issue.

pooling ("revenue pooling"): A telephone industry term meaning setting up special collections of funds for intended cross-subsidy, as in averaging rates between high-cost rural services and less expensive urban ones.

POTS: Jargon for "plain old telephone service."

public switched telephone network: The more formal name given to the commercial telephone business in the United States; includes all the operating companies.

PUC: Public Utility Commission, usually the state entity that sets telephone rates.

pulse code modulation (PCM): A technique by which a signal is sampled periodically, each sample quantitized, and transmitted as a signal binary code.

RBOC: See *regional holding company*.

regional holding company (RHC, RBOC): The companies formed to take over the individual Bell system operating companies at divestiture; there are seven (e.g., Pacific Telesis).

separations: A telephone industry term meaning methods for dividing costs, revenues, etc. between different types of carriers, especially long distance versus local exchanges.

slow-scan television: A technique of placing video signals on a narrowband circuit, such as telephone lines, which results in a picture changing every few seconds.

software: The written instructions which direct a computer program. Any written material or script for use on a communications system or the program produced from the script (see *hardware*).

tariff: The published rate for a service, equipment, or facility established by the communications common carrier.

telco: Jargon for "telephone company."

telecommuting: The use of computers and telecommunications to enable people to work at home. More broadly, the substitution of telecommunications for transportation.

teleconference: The simultaneous visual and/or sound interconnection that allows individuals in two or more locations to see and talk to one another in a long-distance conference arrangement.

telemarketing: A method of marketing that emphasizes the use of the telephone and other telecommunications systems.

teletext: The generic name for a set of systems which transmit alphanumeric and simple graphical information over the broadcast (or one-way cable) signal, using spare line capacity in the signal for display on suitably modified TV receiver.

telex: A dial-up telegraph service.

terminal: A point at which a communication can either leave or enter a communications network.

terminal emulator: Use of a personal computer to act as a dumb terminal; this requires special software or firmware.

timesharing: When a computer can support two or more users. The large computers used by the information utilities can accommodate many users simultaneously who are said to be timesharing on the system.

transponder: The electronic circuit of a satellite that receives a signal from the transmitting earth station, amplifies it, and transmits it to the Earth at a different frequency.

trunk: A main cable that runs from the head end to a local node, then connects to the drop running to a home in a cable television system; a main circuit connected to local central offices with regional or intercity switches in telephone systems.

twisted pair: The term given to the wires that connect local telephone circuits to the telephone central office.

universal service: Traditionally defined as making voice telephone service easily available at affordable cost. In the coming years, it could be broadened to include other telecommunications services.

uplink: The communications link from the transmitting Earth station to the satellite.

upload: To transfer information out of the memory or disk file of your computer to another computer.

videotex: The generic name for a computer system that transmits alphanumeric and simple graphics information over the ordinary telephone line for display on a video monitor.

VLSI: Very large scale integration; single integrated circuits that contain more than 100,000 logic gates on one microchip (see *LSI*).

VSAT: Very small aperature terminal; a special satellite dish especially valuable for networking businesses.

WATS: Wide area telephone service; a service offered by telephone companies in the United States that permits customers to make dial calls to telephones in a specific area for a flat monthly charge, or to receive calls collect at a flat monthly charge.

References

Arms, C., ed. (1988). *Campus Networking Strategies*. Bedford, Mass.: Digital Press.

Augustson, J. G. (1988). "The Pennsylvania State University." In C. Arms, ed., *Campus Networking Strategies*. Bedford, Mass.: Digital.

Aumente, J. (1987). *New Electronic Pathways: Videotex, Teletext, and On-line Data Bases*. Newbury Park, Calif.: Sage.

Baer, P. (1989). "BARRNet." In T. LaQuey, ed., *User's Directory of Computer Networks Accessible to the Texas Higher Education Network Member Institutions*. Bedford, Mass.: Digital.

Barrera, E. (1988). *Advanced Telecommunications between Mexico and Texas: The Maquiladoras*. Austin, Tex.: Center for Research on Communications Technology and Society, The University of Texas at Austin.

Barrera, E., and F. Williams (1990). "Mexico and the United States: The Maquiladora Industries." In F. Williams and David V. Gibson, eds., *Technology Transfer: A Communication Perspective*. Newbury Park, Calif.: Sage.

Bell, D. (1976). *The Coming of Post-Industrial Society*. New York: Basic Books.

Bollier, D. (1988). *The Importance of Communications and Information Systems to Rural Development in the United States*. Report of an Aspen Institute Conference. Turo, Mass.: Aspen Institute for Humanistic Studies.

Borders, W. (1980). "Versatile TV System Helps Britons Buy and Learn." *New York Times*, Section C, p. 19, June 26.

Bozzo, U., and D. V. Gibson (1990). "Italy: Tecnopolis Novus Ortus and the EEC." In F. Williams and D. V. Gibson, eds., *Technology Transfer: A Communication Perspective*. Newbury Park, Calif.: Sage.

Brand, S. (1987). *The Media Lab*. New York: Penguin Books.

Branscomb, A., ed. (1986). *Toward a Law of Global Communications Networks*. New York: Longman.

Brownstein, C. N, (1978). "Interactive Cable TV and Social Services." *Journal of Communications* 28 (2) (Spring): 142–47.

Brunner, M. E. (1986). "Regulation and Rural Telephony." *Telephony*, September 8, 1986.

Castells, M., ed. (1985). *High Technology, Space and Society*. Beverly Hills, Calif.: Sage.

Crenshaw, A. B. (1989). "Prodigy's Bid for Home Banking: Sears-IBM Venture Aims to Computerize a Wide Variety of Services." *Washington Post*, Section H. p. 7, May 28.

Crook, D. (1983). "Times Mirror Plans Videotex News Service." *Los Angeles Times*, Section IV, p. 2, April 22.

Dillman, D. A. (1985). "The Social Impacts of Information Technology in Rural North America." *Rural Sociology* 50 (1): 1–26.

Dillman, D. A., and D. M. Beck (1988). "Information Technologies and Rural Development in the 1990s." *Journal of State Government* 61(1): 29–38.

Dizard, W. P. (1989). *The Coming Information Age: An Overview of Technology, Economics, and Politics*, 3d ed. New York: Longman.

Dordick, H. S. (1990). "The Origins of Universal Service." *Telecommunications Policy* (June) 223–31.

Dordick, H.S., and F. Williams (1986). *Innovative Management Using Telecommunications: A Guide to Opportunities, Strategies, and Applications*. New York: John Wiley.

Dordick, H. S., H. G. Bradley, and B. Nanus, eds. (1980). *The Emerging Network Marketplace*. Norwood, N.J.: Ablex.

Dozier, D. M., and J. A. Ledingham (1982). "Perceived Attributes of Interactive Cable Services among Potential Adopters." Paper presented to the Human Communication Technology Special Interest Group, International Communication Association Annual Convention, Boston, Mass. May.

Drucker, P. (1989). *The New Realities*. New York: Harper and Row.

Dutton, W. H., J. G. Blumler, and K. K. Kraemer, eds. (1986). *Wired Cities: Shaping the Future of Communications*. Boston: G. K. Hall.

Epstein, N. (1986). "Et Voila! Le Minitel." *New York Times Magazine*, p. 46ff, March 9.

Executive Office of the President (1989). *The Federal High Performance Computing Program*. Washington, D.C.: Office of Science and Technology Policy.

Faulhaber, G. R. (1987). *Telecommunications in Turmoil: Technology and Public Policy*. Cambridge, Mass.: Ballinger.

FCC (1985). "Access Charges: MTS and WATS Market Structure." *Federal Register* 50 (72) p. 939, January 8.

Feigenbaum, E. P. M., and H. P. Nii (1988). *The Rise of the Expert Company*. New York: Times Books.

Fernbach, S. (1983). "Scientific Use of Computers." In Michael Dertouzos and J. Moss, eds., *The Computer Age: A Twenty-Year View*. Cambridge, Mass.: MIT Press.

Flamm, K. (1988). *Creating the Computer: Government, Industry, and High Technology*. Washington, D.C.: Brookings Institute.

Glasmeier, A. K. (1991). *Rural America in the Age of High Technology*. New Brunswick, N.J.: Center for Urban Policy Research, Rutgers University Press.

Goldmark, P. C. (1972). "Communication and the Community." In *Communication*. A *Scientific American* Book. San Francisco: W. H. Freeman. See also his "Tomorrow We Will Communicate to Our Jobs." *The Futurist* 6, pp. 35–42, 2 April.

Gordon, K., and J. Haring (1988). *The Effects of Higher Telephone Prices on Universal Service*. FCC Office of Planning and Policy Working Paper Series.

Gore, A. (1989). "National High-Performance Computer Technology Act of 1989." *Newsletter*, May 19.

Greenberger, M., ed. (1985). *Electronic Publishing Plus: Media for a Technological Future*. White Plains, N.Y.: Knowledge Industry.

Hardy, A. P. (1980). "The Role of the Telephone in Economic Development." *Telecommunications Policy* 4: 278–86.

Huber, P. (1987). *Geodesic Network: 1987 Report of Competition in the Telephone Industry*. Washington, D.C.: Government Printing Office.

Hudson, H. E. (1984). *When Telephones Reach the Village*. Norwood, N.J.: Ablex.

———. "Ending the Tyranny of Distance: The Impact of New Communications Technologies in Rural North America." In J. R. Schement and L. Lievrouw, eds., *Competing Visions, Complex Realities: Social Aspects of the Information Society*. Norwood, N.J.: Ablex.

———. (1990). *Communication Satellites: Their Development and Impact*. New York: Free Press.

Johnson, L. L. (1988). *Telephone Assistance Programs for Low-Income Households*. Santa Monica, Calif.: The RAND Corporation.

Katz, E., M. Gurevitch, and H. Hass (1973). "On the Uses of the Mass Media for Important Things." *American Sociological Review* 38: 164–81.

Keller, S. (1977). "The Telephone in New (and Old) Communities." In I. Pool, ed. *The Social Impact of the Telephone*. Cambridge, Mass.: MIT Press.

Kozmetsky, G. (1990). "The Coming Economy." In F. Williams and D. V. Gibson, eds., *Technology Transfer: A Communication Perspective*. Newbury Park, Calif.: Sage.

LaQuey, T. L., ed. (1989). *User's Directory of Computer Networks Accessible to the Texas Higher Education Network Member Institutions*. Bedford, Mass.: Digital.

Lederberg, J., and K. Uncapher (1989). *Towards a National Collaboratory*. Report of an Invitational Workshop at the Rockefeller University.

Lum, P. (1984). "Telephone Use by Senior Citizens: Community Snapshot." Unpublished paper. Annenberg School of Communications, University of Southern California, Los Angeles, Calif.

Magid. L. J. (1989). "On-line Services Leaping Forward." *Los Angeles Times*, Section IV, p. 3, July 13.

Markoff, J. (1989). "Bush Plan Would Aid Computing." *New York Times*, Section D, pp. 1, 3, September 8.

Moss, M. (1986). "Telecommunications and the Economic Development of Cities." In W. H. Dutton, J. G. Blumler, and K. L. Kraemer, eds., *Wired Cities: Shaping the Future of Communications*. Boston, Mass.: G. K. Hall.

Nillis, J., F. R. Carlson, Jr., and G. H. Hanneman (1976). *The Telecommunications Transportation Tradeoff: Options for Tomorrow*. New York: John Wiley.

Noll, R. (1988). *Telecommunications Regulation in the 1990s*. Publication no. 140 Stanford, Calif.: Center for Economic Policy Research, August.

Nora, S., and A. Minc (1980). *The Computerization of Society*. Cambridge, Mass.: MIT Press.

NTIA (1988). *Telecom 2000*. NTIA Special Publication 88–21, Washington, D.C.

NYSERNet, Inc. (1987). *The New York State Education and Research Network*. Troy, N.Y.: NYSERNet, Inc.

NYSERNet News (1989). Vol. 2, no. 8, July/August. Troy N.Y.: NYSERNet, Inc.

Office of Science and Technology Policy (1989). *The Federal High Performance Computing Program*. Executive Office of the President, Washington D.C.

O'Leary, T. J., and B. Williams (1985). *Computers and Information Processing*. Menlo Park, Calif.: Benjamin/Cummings.

Opinion Research Corporation (1983). *Segmentation Study of the Urban/Suburban Cable Television Market*. Prepared for the National Cable Television Association. Princeton, N.J.: Opinion Research Corporation.

OTA (1989). *Linking for Learning*. Washington, D.C.: U.S. Congress, Office of Technology Assessment (Publication SET-439).

———. (1990). *Critical Connections: Communications for the Future*. Washington D.C.: U.S. Congress, Office of Technology Assessment (Publication OTA-CIT-407).

Pacific Bell (c. 1988). *Pacific Bell's Response to the Intelligent Network Task Force Report*. San Francisco: Pacific Bell.

Parker, E. B. (1981). *Economic and Social Benefits of the REA Telephone Loan Program*. Mountainview, Calif.: Equatorial Communications.

Parker, E. B., H. E. Hudson, D. A. Dillman, and A. D. Roscoe (1989). *Rural America in the Information Age: Telecommunications Policy for Rural Development*. Washington, D.C.: The Aspen Institute.

Phillips, A. F., P. Lum, and D. Lawrence (1983). "An Ethnographic Study of Telephone Use." Paper presented to the Fifth Annual Conference on Culture and Communication, Philadelphia, Penn., March.

Pressler, L., and K. V. Schieffer (1988). "A Proposal for Universal Telecommunications Service." *Federal Communications Law Journal* 40 (3), pp. 15–21 (May).

Quarterman, J. S. (1990). *The Matrix: Computer Networks and Conferencing Systems Worldwide*. Bedford, Mass.: Digital.

Roberts, M. (1988). "Introduction." In C. Arms, ed., *Campus Networking Strategies*. Bedford, Mass.: Digital.

Rosengren, K. E., L. A. Wenner, and P. Palmgreen (1985). *Media Gratifications Research*. Beverly Hills, Calif.: Sage.

Saunders, R., J. Warford, and B. Wellenius (1983). *Telecommunications and Economic Development*. Baltimore, Maryland: Johns Hopkins University Press.

Sawhney, H. (1988). "Technology Transfer: An Overview. "In F. Williams, D. Gibson, and H. Sawhney, *Technology Transfer: A Report on Project Development*. Austin, Tex.: Center for Research on Communication Technology and Society, The University of Texas at Austin.

Schmandt, J., F. Williams, R. H. Wilson, and S. Strover, eds. (1990). *The New Urban Infrastructure: A Study of Large Telecommunication Users*. New York: Praeger.

———. (1991). *Telecommunications and Rural Development*: A Study of Business and Public Service Applications, New York: Praeger.

Schmandt, J., F. Williams, and R. H. Wilson, eds. (1989). *Telecommunication Policy and Economic Development: The New State Role*. New York: Praeger.

Shaver, T. L. (1983). "The Uses of Cable TV." Unpublished Master's thesis, University of Kentucky.

Smith, R. L. (1972). *The Wired Nation: Cable TV: The Electronic Communications Highway*. New York: Harper and Row. See also: "The Wired Nation." *The Nation*, May 18, 1970.

Stoffregen, P. E. (1988). *Telecommunications Deregulation: A State-by-State Analysis of Legislation*. Alexandria, Va.: Telecom Publishing Group.

Strassman, P. A. (1985). *Information Payoff*. New York: Free Press.

Strover, S. (1987). "Urban Policy and Telecommunications." *Urban Affairs Quarterly* 10 (4): 341–56.

———. (1988). "Urban Telecommunication Investment." In Frederick Williams, ed., *Measuring the Information Society*. Newbury Park, Calif.: Sage.

Strover, S., and F. Williams (1990). *Rural Revitalization and Information Technologies in the United States*. Report to the Ford Foundation.

Tatsuno, S. (1986). *The Technopolis Strategy*. New York: Prentice-Hall.

Tempest, R. (1989). "Minitel: Miracle or Monster?" *Los Angeles Times*, Section A., p. 1, October 24.

U.S. Department of Agriculture (1987). *Rural Economic Development in the 1980s, A Summary*. Agriculture Information Bulletin Number 533, October.

Wall Street Journal (1982). "Knight-Ridder Plans Home-Data Venture with A. S. Able," Section II, p. 30 (W), July 27.

Williams, F. (1983). *The Communications Revolution*. New York: New American Library.

————. (1988d). *The Competitive Challenge: Interexchange Carrier and State Telecommunications Policy*. Austin, Tex.: Center for Research on Communication Technology and Society, The University of Texas at Austin.

————. (1988b). *Telecommunications and Rural Development in the United States: A Policy View*. Research Report. Austin, Tex.: Center for Research on Communications Technology and Society, The University of Texas at Austin.

————. (1988c). *Telecommunications as Strategic Investment*. Research Report. Austin, Tex.: Center for Research on Communication Technology and Society, The University of Texas at Austin.

————. (1991). "The Intelligent Network: A New Beginning for Information Services on the Public Network?" In F. Phillips, ed., *Thinkwork: Working, Learning and Managing in a Computer-Interactive Society*. New York: Praeger.

Williams, F., and E. Brackenridge (1990). *Transfer via Telecommunications: Networking Scientists and Industry*. Newbury Park, Calif.: Sage.

Williams, F., and H. S. Dordick (1983). *The Executive's Guide to Information Technology*. New York: John Wiley.

Williams, F., and D. V. Gibson, eds. (1990). *Technology Transfer: A Communication Perspective*. Newbury Park, Calif.: Sage.

Williams, F. and S. Hadden (1991). "On the Prospects for Redefining Universal Service: From Connectivity to Content." In J. Schement and B. Ruben, eds., *Information and Behavior*, vol. 5. New Brunswick, N.J.: Transaction.

Williams, F., H. S. Dordick, and N. Contractor (1986). "Predicting Reported Telephone Usage. "In F. Williams, H. S. Dordick, and Associates, *Social Research and the Telephone*. Research Report. Los Angeles: Annenberg School of Communications at the University of Southern California.

Williams, F., R. E. Rice, and E. M. Rogers (1988). *Research Methods and New Media*. New York: Free Press.

Williams, F., H. Sawhney, and E. Brackenridge (1990). *Rural Telecommunications and Development: A Study of Customer and Applications at Selected Ohio Sites*. Austin, Texas: Report to the United Telephone Co. of Ohio.

Wulf, W. A. (1989). "Government's Role in the National Network." *Educom Review* 24 (2) (Summer): 22–26.

Wurtzel, A. H., and C. Turner (1977). "What Missing the Telephone Means." *Journal of Communication* 27:48–56.

Zuboff, S. (1988). *In the Age of the Smart Machine*. New York: Basic Books.

Index

237

About the Author

Professor Frederick Williams occupies the Mary Gibbs Jones Centennial Chair in Communications at the University of Texas at Austin, where he directs the Center for Research on Communication Technology and Society (one of the University's 80 organized research centers). He is also the W. W. Heath Centennial Research Fellow in the IC2 Institute, a business innovation research center at the university, and a Visiting Professor in the Lyndon Baines Johnson School of Public Affairs. Most recently, he was appointed as a Senior Fellow in the Gannett Center for Media Studies at Columbia University where he served in residence in 1991.

Fred and his wife, Dr. Victoria Williams, and family divide their time between homes in Fairlawn, New Jersey and Austin, Texas.